Power Electronics for Photovoltaic Power Systems

Synthesis Lectures on Power Electronics

Editor
Jerry Hudgins, *University of Nebraska-Lincoln*

Synthesis Lectures on Power Electronics will publish 50- to 100-page publications on topics related to power electronics, ancillary components, packaging and integration, electric machines and their drive systems, as well as related subjects such as EMI and power quality. Each lecture develops a particular topic with the requisite introductory material and progresses to more advanced subject matter such that a comprehensive body of knowledge is encompassed. Simulation and modeling techniques and examples are included where applicable. The authors selected to write the lectures are leading experts on each subject who have extensive backgrounds in the theory, design, and implementation of power electronics, and electric machines and drives.

The series is designed to meet the demands of modern engineers, technologists, and engineering managers who face the increased electrification and proliferation of power processing systems into all aspects of electrical engineering applications and must learn to design, incorporate, or maintain these systems.

Power Electronics for Photovoltaic Power Systems
Mahinda Vilathgamuwa, Dulika Nayanasiri, and Shantha Gamini
2015

Digital Control in Power Electronics, 2nd Edition
Simone Buso and Paolo Mattavelli
2015

Transient Electro-Thermal Modeling of Bipolar Power Semiconductor Devices
Tanya Kirilova Gachovska, Bin Du, Jerry L. Hudgins, and Enrico Santi
2013

Modeling Bipolar Power Semiconductor Devices
Tanya K. Gachovska, Jerry L. Hudgins, Enrico Santi, Angus Bryant, and Patrick R. Palmer
2013

Signal Processing for Solar Array Monitoring, Fault Detection, and Optimization
Mahesh Banavar, Henry Braun, Santoshi Tejasri Buddha, Venkatachalam Krishnan, Andreas Spanias, Shinichi Takada, Toru Takehara, Cihan Tepedelenlioglu, and Ted Yeider
2012

The Smart Grid: Adapting the Power System to New Challenges
Math H.J. Bollen
2011

Digital Control in Power Electronics
Simone Buso and Paolo Mattavelli
2006

Power Electronics for Modern Wind Turbines
Frede Blaabjerg and Zhe Chen
2006

Power Electronics for Photovoltaic Power Systems

Mahinda Vilathgamuwa, Dulika Nayanasiri, and Shantha Gamini

ISBN: 978-3-031-01372-0   paperback
ISBN: 978-3-031-02500-6   ebook

DOI 10.1007/978-3-031-02500-6

A Publication in the Springer series
SYNTHESIS LECTURES ON POWER ELECTRONICS

Lecture #8
Series Editor: Jerry Hudgins, *University of Nebraska-Lincoln*
Series ISSN
Print 1931-9525   Electronic 1931-9533

Power Electronics for Photovoltaic Power Systems

Mahinda Vilathgamuwa
Queensland University of Technology

Dulika Nayanasiri
University of Moratuwa

Shantha Gamini
Australian Maritime College, University of Tasmania

SYNTHESIS LECTURES ON POWER ELECTRONICS #8

ABSTRACT

The world energy demand has been increasing in a rapid manner with the increase of population and rising standard of living. The world population has nearly doubled in the last 40 years from 3.7 billion people to the present 7 billion people. It is anticipated that world population will grow towards 8 billion around 2030. Furthermore, the conventional fossil fuel supplies become unsustainable as the energy demand in emerging big economies such as China and India would rise tremendously where the China will increase its energy demand by 75% and India by 100% in the next 25 years. With dwindling natural resources, many countries throughout the world have increasingly invested in renewable resources such as photovoltaics (PV) and wind.

The world has seen immense growth in global photovoltaic power generation over the last few decades. For example, in Australia, renewable resources represented nearly 15% of total power generation in 2013. Among renewable resources, solar and wind account for 38% of generation. In near future, energy in the domestic and industrial sector will become "ubiquitous" where consumers would have multiple sources to get their energy. Another such prediction is that co-location of solar and electrical storage will see a rapid growth in global domestic and industrial sectors; conventional power companies, which dominate the electricity market, will face increasing challenges in maintaining their incumbent business models.

The efficiency, reliability and cost-effectiveness of the power converters used to interface PV panels to the mains grid and other types of off-grid loads are of major concern in the process of system design. This book describes state-of-the-art power electronic converter topologies used in various PV power conversion schemes. This book aims to provide a reader with a wide variety of topologies applied in different circumstances so that the reader would be able to make an educated choice for a given application.

KEYWORDS

active power decoupling, centralized PV power conversion, distributed PV power conversion, energy storage interfacing, isolated DC-DC converters, multi-level converters, non-isolated DC-DC converters, photovoltaic power systems, power converter control, power electronics, soft-switching, micro inverters, micro converters, module integrated converters

Contents

CHAPTER 1

PV Power Conversion Systems

1.1 INTRODUCTION

The term "photovoltaic" means the direct conversion of light energy to electrical energy by means of photovoltaic (PV) cells. According to historical records, in 1839, A.E. Becquerel experimented with a "wet cell" and recorded the first "photo-electric" effect where the current flown out of the wet cell increased with the intensity of light shone on to the cell. In 1877, a paper published in *Proceedings of the Royal Society, London,* by Adams and Day, reported photovoltaic effect of the selenium cell they built. However, it was Albert Einstein who made a significant breakthrough on the understanding of photovoltaic effect in 1904 by publishing his work on quantum theory. The practical silicon-based PV cell was built by Pearson, Chapin, and Fuller in Bell laboratories in 1954. The doped-silicon photovoltaic cell built had 6% efficiency. PV cells at that time used in special applications due to their high manufacturing cost. However, subsequently they were adopted in space satellites in a major way since their installation in U.S. satellite Vanguard 1 in 1958. Although their initial terrestrial applications were limited, with energy crisis that took place in 1970s and nuclear power station disasters in Harrisburg and Chernobyl, major industrialized nations decided to pay increasing attention in developing alternate energy sources such as PV. With the decline in production cost, improved efficiency and with the introduction of feed-in tariff system by major industrialized countries such as Germany and Italy for single family homes, the photovoltaic power received a major boost. Apart from household PV generations, there are large numbers of PV power stations installed throughout the world.

PV cells have gained popularity and high growth rate within the last decade due to several reasons compared to other renewable energy sources. Among them, low greenhouse gas emission during life time of PV systems is a significant factor for PV energy being favored and as such giving it a larger share in energy mix in a number of European countries and the U.S. Furthermore, energy payback time (time taken by PV systems to pay back cumulative energy demand arise throughout the life time of the system) of PV systems is reduced to a value as low as 1–2 years depending on the geographical location [1]. Another most important factor related to PV cell is its grey energy which is defined as the energy that goes into the product and its raw material. The grey energy of the PV cell is significantly reduced with thin wafer technologies (due to reduced amount of raw material requirement) and efficient production technologies.

The development of PV cells technologies can be divided in to three major stages which are known as first, second and third generations. First generation PV cells are based on silicon wafers and they typically exhibit performance around 15–20%. PV cells belong to this generation have shown stable operation in addition to satisfactory performance. However, energy consumption to manufacture these PV cells is significantly high. The second generation PV cells are based on amorphous silicon, Cadmium Telluride (CdTe), and Copper Indium Gallium Selenide (CIGS). These PV cells require a large input energy at the production stage. But they need a relatively low amount of raw material with the absence of wafer technology. The power conversion efficiency of these PV cells ranges from 10–15% [2] and these cells printed on thin films and are flexible compared to thin silicon crystal-based first-generation solar cells [3]. The third generation solar cells are based on new technologies such as organic PV (OPV) [4], dye-sensitized solar cells (DSSC) [5], multi-junction solar cells [6] which show record breaking energy conversion efficiency in the laboratory tests, and other types of solar cells that are based on quantum dots [7]. The main objectives behind developing third generation solar cells are to reduce production cost and material usage, to have quick and inexpensive large scale production and to reduce adverse environmental effects by minimizing waste during production. The advancement of cell printing technologies, and their stability and efficiency play key roles in making OPV and DSSC as the most promising technologies based on current status of third-generation PV cell developments.

The sun, which is the key source of energy in photovoltaic systems can be considered as a huge nuclear fusion reactor that produces 3.89×10^{20} MW of power having approximately $6000°C$ surface temperature. Sun radiates its energy in a wide spectrum and the earth's atmosphere receives around 1370 W/m^2 of sun's energy. Nevertheless, all such energy reached at the earth's atmosphere cannot be received by the earth's surface as atmospheric gases and water vapor attenuates solar radiation while around one third is reflected back to the outer space and another significant portion is scattered. Therefore, the amount of solar energy received at the earth's surface is dependent on the "Air mass" (AM) of the atmosphere as given by (1.1). AM0 represents the solar radiation level at the outer periphery of the atmosphere where there are no atmospheric gases present, as shown in Fig. 1.1. On the other hand, AM1 represents the solar radiation received when the sun is directly overhead (at zenith). If the sun's angle is θ from its zenith as shown

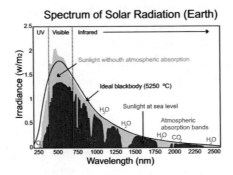

Figure 1.1: Spectrum of solar radiation.

in Fig. 1.2, the air mass index is defined as follows

$$AM = \frac{xa}{xb} = \frac{1}{\cos \theta},$$

(1.1)

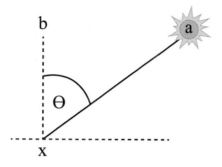

Figure 1.2: Air mass index.

Standard test conditions to measure PV cell performance are at 25°C and at AM1.5 where sun's zenith angle is 48.19° and total solar irradiance received under such conditions is approximately 1000 W/m^2.

1.2 PRINCIPLES OF PV CELL OPERATION

PV cells are manufactured using semiconductor materials such as Si, GaAs, and CdTe and they are classified in between insulators and conductors in terms of their electrical conductivity. According to the quantum theory, energy bands can be defined for intrinsic semiconductor materials as shown in Fig. 1.3a. The electrons in the conduction band contribute for electrical conductivity. Therefore, as the name suggests, conductors have large number of electrons in their conduction band while insulators hardly possess any electrons in their conduction band. On the other hand, semiconductor materials have electrical conductivity between that of conductors and insulators.

In absolute zero temperature, there are no electrons in the conduction band of semiconductor materials. At room temperature, however, a few electrons can be elevated into the conduction band with the absorption of some thermal energy. The energy band diagram for a normal (intrinsic) semiconductor at room temperature is shown in Fig. 1.3a. Moreover, electrons in the valence band can jump into conduction band if they receive enough energy from photons which is greater than the band gap energy (E_g) of the semiconductor material when it is exposed to sunlight or artificial light. For example, Si has a band gap E_g of 1.12 eV while GaAs has a band gap of 1.4 eV. As such, with the absorption of light energy electrons get enough energy to jump into the conduction band. Figure 1.3b depicts such a situation where an electron jumps to the conduction band creating an absence of an electron in the valance band. This absence of an electron is

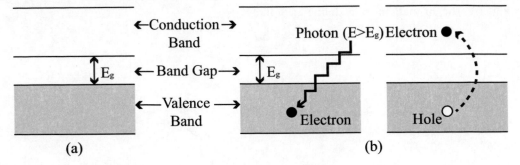

Figure 1.3: Energy band diagram for a normal (intrinsic) semiconductor: (a) at room temperature and (b) the migration of an electron upon absorbing energy from a photon ($> E_g$).

referred to as a hole. Eventually, this creates an electron-hole pairs in the semiconductor material and free electrons and holes wander around within the material for some time before they are recombined without producing any useful electrical effect. In order to get such free electrons and holes to produce any electricity a potential barrier within the semiconductor material must be set up. In order to create this potential barrier, two semiconductors with different characteristic are put together and a junction is formed which is known as a p-n junction.

1.3   P-N JUNCTION

Semiconductor materials such as silicon have 4 valence electrons in their outer shell. Silicon can be doped by adding 5th and 3rd column elements in the periodic table to produce n-type and p-type semiconductor materials, respectively. When doped with 5th column material such as phosphorous (P), silicon atoms form a crystalline structure with the added impurity. In the formation of crystalline structure, 4 of the 5 outer shell electrons of phosphorous are used for bonding and one valence electron is not used. These unused electrons can get loose easily with the absorption of thermal energy and thus creating extra electrons with improved conductivity. With the availability of extra electrons for conduction such semiconductor materials are called n-type semiconductor materials.

On the other hand, if silicon is doped with a 3rd column material such as boron (B) with 3 valence electrons, a crystalline structure is formed with boron occupying the place of some silicon atoms. In such a situation, a "hole" in the bond with neighboring silicon atoms exists and abundance of such holes make these materials p-type semiconductor materials with increased conductivity.

By interfacing n-type and p-type semiconductor materials, a simple p-n junction can be formed and that becomes the building block of photovoltaic power generation. Electrons in the n-type semiconductor tend to diffuse through the interface into p-type side by leaving resultant charge behind. Similarly, holes in p-type material would migrate to n-type material by leaving negative charge in p-type material. This charge movement is shown in Fig. 1.4.

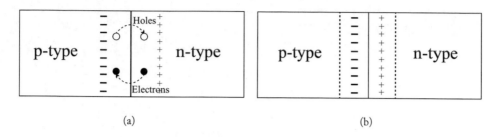

(a)   (b)

Figure 1.4: (a) Simple p-n junction and (b) formation of depletion layer.

In equilibrium, due to unbalance in charges in both p-type and n-type semiconductor materials an electric field is set up which in turn prevents any further movement of majority carriers across the interface. This phenomena result in depletion of free carriers around the interface and formation of a depletion layer as shown in Fig. 1.4b. In this situation, only a minute amount of thermally agitated minority carriers can flow across the junction and the flow of current is known as reverse leakage current.

The aforementioned description is valid for any general p-n junction. Moreover, the p-n junction is known as diode as well. The depletion layer can be enlarged if a reverse voltage applied across the junction and alternatively it can be diminished with any applied forward voltage. With the application of forward voltage, the electric field causing depletion layer vanishes and a forward electric field that enables majority carrier movement is set up. The schematic of a p-n junction is shown in Fig. 1.5a and its symbolic form is shown in Fig. 1.5b. The ideal p-n junction or diode voltage-current characteristics can be described using Shockley diode equation as in (1.2) and can be shown graphically, as in Fig. 1.5c. The forward current would exponentially rise with the increase of the forward voltage as shown in Fig. 1.5c:

$$I_d = I_0 \left(e^{\frac{qV_d}{kT}} - 1 \right). \tag{1.2}$$

In Eq. (1.2), I_d represents the diode current (A), V_d is the voltage across the diode (V), I_0 is the diode leakage current (A) (or saturation current under dark conditions as applicable to PV cells), q is the charge of an electron (1.602×10^{-19} C), k is Boltzman's constant (1.381×10^{-23} J/K) and T is the junction temperature (K).

1.4   THE PHOTOVOLTAIC EFFECT

As described earlier when a sufficiently strong light source is shone on a PV cell, light energy causes the creation of electron-hole pairs within the PV cell. Once rays of light arrive on the PV

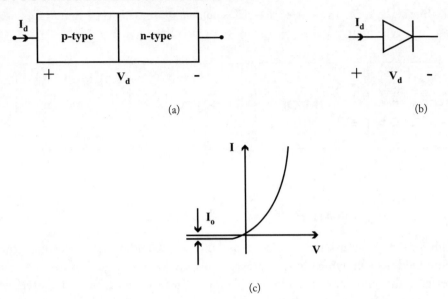

Figure 1.5: (a) p-n junction with its current and voltage indicated, (b) symbolic notation of p-n junction (diode), and (c) V-I characteristics of the p-n junction.

cell they can either be absorbed or reflected from the top or bottom surface of the cell. Those photons of light absorbed either have low strength, and they just cause atomic vibrations and get dissipated as heat, or excessive strength, that enable them to create electron-hole pair and the rest of energy get dissipated as heat. Those photo-generated electron-hole pairs must be separated and collected at their respective electrodes if PV cell can be used as an electrical generator. Previously described electric field across the depletion layer in the PV cell plays an important role in doing just that. If photo-generated electron-hole pairs appear in either p- or n-side of the PV cell, those electrons are quickly swept towards the n-side of the cell under the influence of the field before they are recombined and similarly holes are attracted towards the p-side of the cell, as shown in Fig. 1.6.

When electrons and holes get accumulated at the top layer of n-type material and at the bottom of the p-type material, respectively, those layers can be connected electrically as shown in Fig. 1.6. Then there would be a flow of electrons originated from the top metal contact towards the back metal contact where they are recombined with the accumulated holes thus creating a current flow. The depletion layer would remain intact as the charge imbalance remains at the junction enabling continual separation of photo-generated electron-hole pairs.

The photo-generated short-circuit current I_{sc} is superimposed with the dark-condition saturated current (leakage current in a p-n junction) as shown in PV voltage-current (V-I) characteristics, in Fig. 1.7.

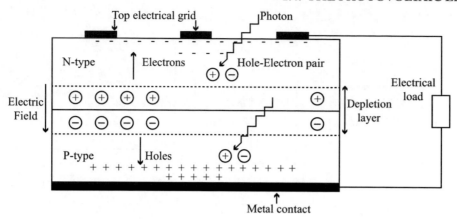

Figure 1.6: Principle of the operation of a photovoltaic cell.

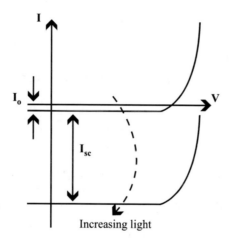

Figure 1.7: PV V-I characteristics.

Mathematically, therefore the voltage-current (V-I) relationship for a PV cell with photo-generated current can be expressed as in (1.3):

$$I = I_0 \left(e^{\frac{qV_d}{kT}} - 1 \right) - I_{sc}.$$
(1.3)

The stronger the light source, the higher would be the photo-generated short circuit current and the voltage-current curve would move further downwards. The most rudimentary electrical equivalent circuit of a PV cell can be drawn as shown in Fig. 1.8a.

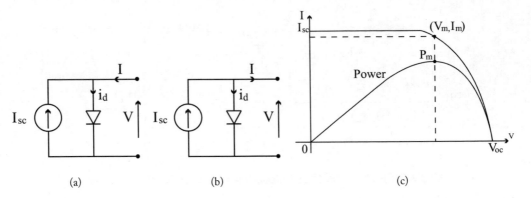

Figure 1.8: (a) PV cell equivalent circuit with current direction indicating it as a passive device, (b) PV cell equivalent circuit with current direction indicating it as an active source, and (c) V-I and V-P characteristics for the circuit shown in (b).

If PV cell is to be represented as an electric source its output current direction needs to be reversed as shown in Fig. 1.8b and its modified voltage-current and voltage-power graphs are then shown in Fig. 1.8c. Then, from (1.3), the modified voltage-current relationship can be obtained as in (1.4):

$$I = I_{sc} - I_0 \left(e^{\frac{qV_d}{kT}} - 1 \right). \tag{1.4}$$

Some important parameters of PV cells such as short-circuit current I_{sc}, open circuit voltage V_{oc} and the maximum power P_m and its corresponding voltage-current coordinates (V_m, I_m), are also shown in Fig. 1.8c. I_{sc} is the photo-generated current and is found to be proportional to the solar insolation and the operating temperature of the solar cell as given in (1.5) [8]:

$$I_{sc} = (I_{sc,n} + K_I \Delta T) \frac{G}{G_n}. \tag{1.5}$$

In (1.5), $I_{sc,n}$ (A) represents the photo-generated current under nominal conditions (25°C and 1000 W/m²), $\Delta T = (T - T_n)$ (T and T_n represent actual and nominal temperatures in K, respectively), G_n is the nominal insolation level, G is the actual insolation level in W/m², and K_I is a constant. Current-voltage and power-voltage characteristics of PV cells for varied insolation levels is shown in Fig. 1.9.

Equation (1.4) represents most general and simplified mathematical model of a solar PV cell. However, cell behavior is too complex to represent comprehensively in mathematical terms. Therefore, without loss of generality, the circuit can be modified to represent two major extrinsic effects using lumped parameter of series and parallel resistances. Although these effects are distributed throughout the cell in nature, lumped parameter-based equivalent circuit provides a

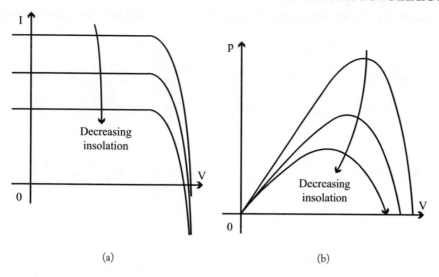

(a)

(b)

Figure 1.9: Photovoltaic cell characteristics with varied insolation: (a) V-I characteristic and (b) V-P characteristic.

realistic way to analyse PV circuits in almost all practical applications. The series resistance (R_s) represents the effects of the resistance of the metallic grid on the top surface of the cell, the resistance of the bulk semiconductor device itself and the resistance of metallic contacts [9]. On the other hand, the parallel resistance (R_p) represents the effects of the leakage around the edge of the cell, small metallic short circuits and diffusion paths along grain boundaries [9]. With the inclusion of R_p and R_s in the equivalent circuit, it can be drawn as in Fig. 1.10.

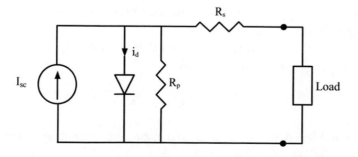

Figure 1.10: PV cell equivalent circuit considering cell series and parallel resistances.

By taking R_s and R_p into consideration, the modified voltage-current characteristic of the solar cell is given as in (1.6):

$$I = I_{sc} - I_0 \left(e^{\frac{qV_d}{kT}} - 1 \right) - \left(\frac{V + IR_s}{R_p} \right). \tag{1.6}$$

Their effect on current-voltage characteristics is shown graphically in Fig. 1.11.

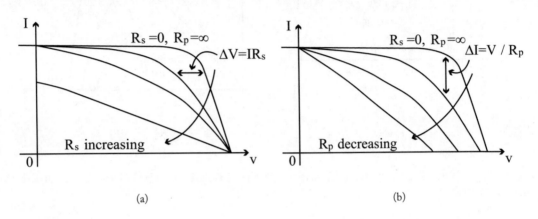

Figure 1.11: PV cell V-I characteristics, considering the effect of (a) R_s and (b) R_p.

1.5   MODULARIZATION OF PV CELLS

Generally, a PV cell produces 0.6 V of voltage and a maximum current of 30 mA/cm^2 under peak insolation level of 1 kW/m^2. Therefore, it is imperative that they are connected in series to obtain a reasonable output voltage required for most practical applications. Such series connected units are known as PV modules and they are shown in schematic form in Fig. 1.12a. PV modules with 36 cells are commonly available in market. If a certain application requires higher voltage, then such modules can be connected in series. These formations are known as solar arrays and are schematically shown in Fig. 1.12b.

Assuming identical cell characteristics, a module with m cells connected in series would have the following voltage-current characteristic:

$$I_{series} = I_{sc} - I_0 \left(e^{\frac{qV_{series}}{kTm}} - 1 \right). \tag{1.7}$$

Alternatively, for a module with n cells connected in parallel, the voltage current characteristic is given by (1.8):

$$I_{parallel} = nI_{sc} - nI_0 \left(e^{\frac{qV_{parallel}}{kT}} - 1 \right). \tag{1.8}$$

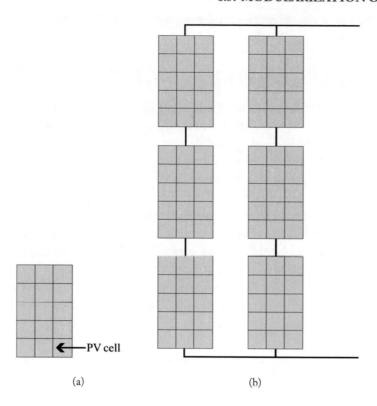

(a)   (b)

Figure 1.12: (a) PV module and (b) PV array.

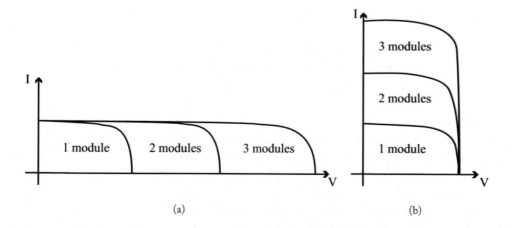

(a)   (b)

Figure 1.13: V-I characteristics when modules are connected in (a) series and (b) parallel.

Similar voltage-current characteristics can be obtained when modules are connected in series or in parallel and are shown graphically in Fig. 1.13.

1.6   BYPASS AND BLOCKING DIODES

Suppose some cells in series connected module get shaded or cracked then they would not be able to produce the same current as their series connected neighbors would produce. In such a situation those cells that are affected by shading or cracking would act as a load instead of a source. Such malfunctioning of faulty cells would create hot spots in the module and module can become permanently damaged if preventive mechanisms are not in place. To overcome this, a bypass diode can be connected in parallel with each cell. Therefore, in situations where a particular cell happens to underperform in comparison to its neighboring cells, the bypass diode would get activated to bypass the underperforming cell as shown in Fig. 1.14a. Under normal conditions, the bypass diode would get deactivated by the induced emf of 0.6 V of the functioning cell, as shown in Fig. 1.14b.

However, it does not make any economical sense if every PV cell gets a bypass diode connected, therefore, a diode is connected across a number of cells that are grouped together as shown in Fig. 1.14c. If one or more number of that group of cells get mal-functioned the bypass diode would get activated.

When parallel PV arrays are in use, a malfunctioning or underperforming string in the array tends to absorb power rather than generating. In such a situation, a blocking diode would prevent reverse current and protect the malfunctioning string, as shown in Fig. 1.15. However, under normal conditions, blocking diodes conduct generated current and give rise to a certain energy loss in day time.

1.7   PHOTOVOLTAIC POWER CONVERSION SYSTEMS

PV power conversion systems can be categorized in to two groups based on their load. The load of PV power conversion systems can be either grid connection or local load (linear or nonlinear). These systems can be further categorized based on the availability of energy storage systems which plays a vital role due to intermittence nature of solar irradiance. The energy storage can be realized using different technologies such as, battery, flywheel, and super capacitors. Super capacitors and flywheels can be used to fulfill short-term energy storage requirements due to their high power density and batteries can be incorporated in systems that need long-term energy storage requirements due to their high energy density. Hence, batteries play a significant role in grid isolated systems. A few such PV system configurations are shown in Fig. 1.16.

Figure 1.16a shows the basic configuration of a grid isolated PV power conversion system with energy storage. Charge controller of the system manages power generated by the PV array based on power demand from the local load and battery charge status. Figures 1.16b and 1.16c show two different configurations of grid-connected PV systems with energy storage. The en-

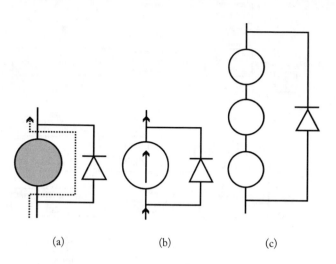

(a)   (b)   (c)

Figure 1.14: The operation of bypass diode when (a) cell is malfunctioning, (b) cell is normal, and (c) economical way of connecting the bypass diode.

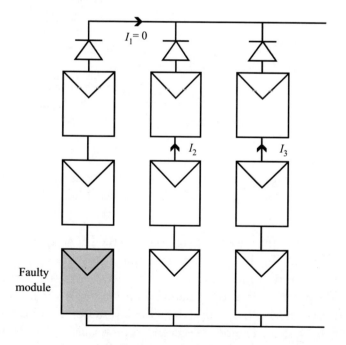

Figure 1.15: Connection of blocking diodes in PV arrays.

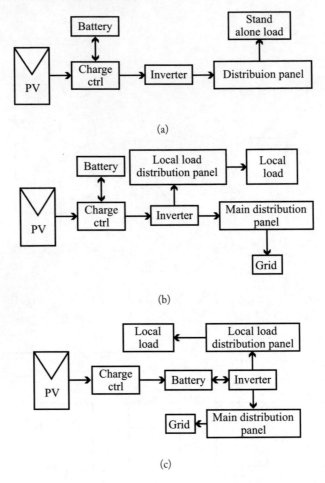

Figure 1.16: Basic configurations of PV power conversion systems with energy storage and (a) grid-isolated, (b) grid interfaced, and (c) grid interfaced with a local load.

ergy storage and inverter of the system are connected to the PV array through a charge controller of the system as shown in Fig. 1.16b. The inverter and energy storage elements can be operated separately. The grid connection and local loads are connected to the inverter through separate distribution panels. The other feasible configuration shown in Fig. 1.16c has a different architecture compared to configuration shown in Fig. 1.16b. The inverter or local load draws power through energy storage of the system. Hence, PV-generated power always is transferred to the energy storage of the system and then is transferred to the load. As a result, efficiency of the system gets significantly affected by the integration of the energy storage to the system. The energy storage not only influences the efficiency of the system, but also the reliability, maintenance, and cost.

However, due to economic and reliability reasons, the majority of present day PV systems are connected to the utility grid directly without any storage, as shown in Fig. 1.17. In such PV power conversion systems, the PV array is directly connected to the power converter and it is in turn connected to the utility grid via a main distribution panel. Some linear or nonlinear local loads can also be connected to the output of the power converter via the main distribution panel. A detailed discussion on grid-connected PV converters and their control is presented in the next section.

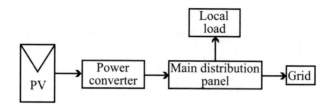

Figure 1.17: Grid-connected PV power conversion system without energy storage.

1.8   GRID INTEGRATION OF PV SYSTEMS

It is necessary to convert PV generated energy into a form that is appropriate for mains grids or for consumer utilization in order to consume it in a useful manner. Such a conversion is carried out by various types of power electronic converters and among them are DC/AC, DC/DC, AC/DC and AC/AC type converters. Typical power converter configurations for grid integration of PV modules are shown in Fig. 1.18.

Centralized inverters shown in Fig. 1.18a are generally used in higher power rating PV power systems which are economical and efficient but suffer from low reliability since there is only one grid connecting inverter for large number of PV modules. If this inverter fails, the whole PV system will be disconnected from the grid. Furthermore, they would not be able to carry out maximum power point tracking at modular level, therefore, optimized generation of PV power would not be possible particularly when some modules in the array are shaded or faulty. On the other hand, modularized systems shown in Figs. 1.18b and 1.18c would act optimally in terms of maximum power point tracking (MPPT) at modular level even though they can be somewhat expensive and lossy. Furthermore, modular structures tend to increase the reliability of the system since a faulty converter or PV module can be disconnected from the system while the others can continue to operate.

Modern day power converters not only convert PV power from one form to another but also they perform numerous ancillary services such as harmonic compensation, reactive power compensation, unbalance compensation, load balancing, and flicker mitigation [10]. The controller used in controlling power converters perform numerous tasks. Among them are power flow con-

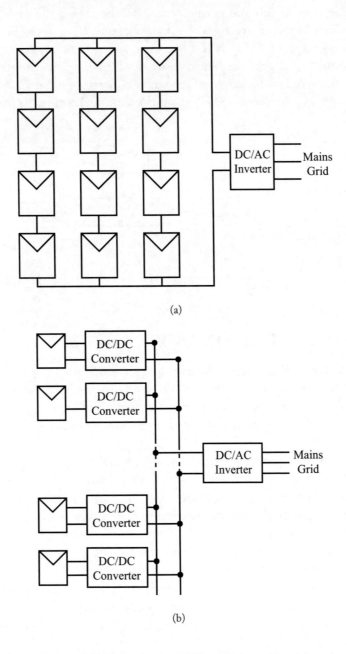

(a)

(b)

Figure 1.18: Grid integration configurations of PV modules using (a) centralized inverters and (b) modularized system with DC-DC converters. *(Continues.)*

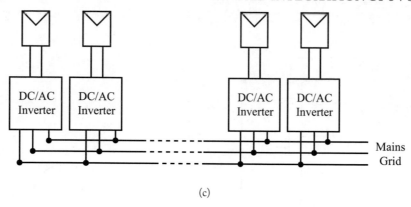

(c)

Figure 1.18: *(Continued.)* Grid integration configurations of PV modules using (c) modularized system with DC-AC inverters.

trol, maximum power point tracking, anti-island detection, fault ride through, and management of energy storage systems [11].

In PV systems, the generated power is a nonlinear function with voltage and its value would vary with the insolation level, as shown in Figs. 1.9a and 1.9b. Therefore, it would be imperative to track the maximum power continually and commonly used maximum power point tracking (MPPT) methods are perturb and observe and incremental conduction methods [12]. The voltage-power (V-P) characteristics are shown in Fig. 1.19b when some PV modules are shaded or receive uneven insolation as shown in Fig. 1.19a. Such situations require more sophisticated MPPT methods in order to track the global maximum power point instead of local power peaks in V-P characteristics curve.

Modern PV power converters are capable of operating in both grid-connected and standalone modes. In the grid-connected mode, as the grid voltage provides voltage and angle reference through a phase-locked-loop, the power converter's basic functionality is transferring maximum power to the grid by using the current control, as shown in Fig. 1.20a. However, if PV converter system is used in a dispatchable power system, an energy storage port is needed as shown in Fig. 1.20b as the PV generated energy tends to be intermittent. More information on the control and other aspects of the dispatchable mode of operation can be found in Chapter 5. In standalone mode, the inverter voltage magnitude and angle are controlled as shown in Fig. 1.20c.

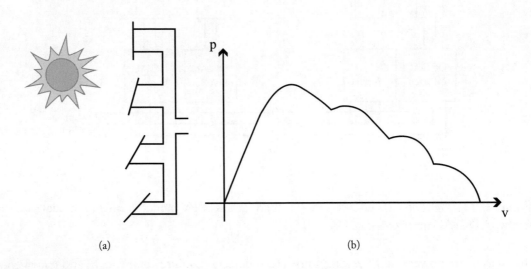

Figure 1.19: (a) PV array with uneven insolation and (b) its P-V characteristics.

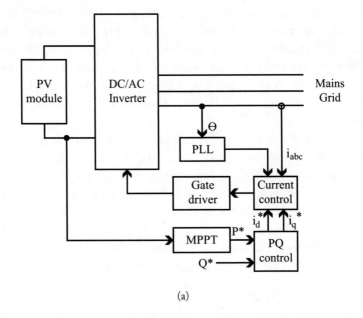

(a)

Figure 1.20: Schematic diagrams of (a) current-controlled PV system. *(Continues.)*

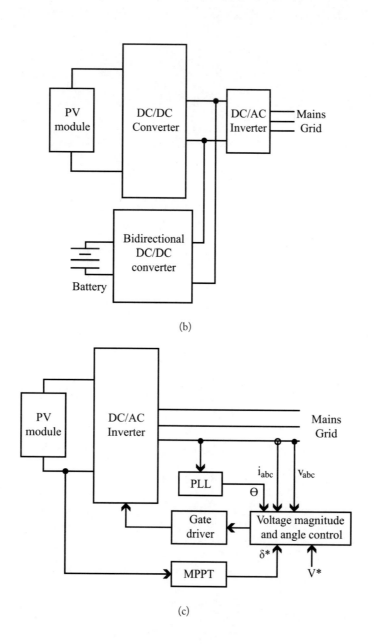

(b)

(c)

Figure 1.20: *(Continued.)* Schematic diagrams of (a) current-controlled PV system, (b) dispatchable PV system with storage, and (c) voltage-controlled PV system.

CHAPTER 2

Centralized PV Power Conversion Systems

2.1 INTRODUCTION

PV power conversion system architectures can be categorized as centralized and distributed based on the interconnection of PV modules and the way MPPT algorithms are implemented. As the name implies, in centralized PV power conversion systems, several PV modules are connected in series and parallel into a single grid-interfacing converter. Instead of individual MPPs of PV modules, a global MPP is tracked in these systems with the use of a central converter. As opposed to the centralized architecture, distributed architecture-based systems use multiple converters to integrate PV modules into the power grid. This enables individual MPP tracking. Based on the series and parallel connection of PV modules into the central inverter, centralized systems can be further subdivided into three categories as central, string, and multi-string. These three configurations are graphically illustrated in Fig. 2.1.

In the central configuration, parallel connected strings comprising series connected PV modules are connected to the DC-side of the central inverter, as shown in Fig. 2.1a. PV modules of the array may not have identical operating characteristics due to several factors such as manufacturing differences as a result of mass production and varied device characteristics (e.g., bypass diode) with aging and external factors such as dirt and soiling. Moreover, it is a well-known fact that temperature of a PV module has a significant effect on its characteristic curve. Temperature distribution of a large PV array may not be even with variable wind patterns across PV array and other factors such as ground conditions. As a result, PV characteristic curve of whole PV array may have multiple peak power points. Nevertheless, the central inverter tracks only a global MPP which might be different from individual MPPs. Therefore, PV modules of the array might not perform optimally and thus the overall efficiency might be lower than expected. The string configuration, shown in Fig. 2.1c, can be used as a solution to this issue where separate DC-AC converters are used for integrating each PV string to the utility grid. This configuration is also known as string inverter-based systems and can be rated up to several kilowatts.

The operation of the string inverter is bit complex as MPP and other functionalities are performed by a single inverter. This drawback can be overcome using multi-string architectures, as shown in Fig. 2.1b. In this system, MPP of the each PV string is tracked using a separate DC-DC converter to obtain better energy conversion efficiency compared to the other two configurations. Moreover, the multi-string configuration can be effectively used to integrate several

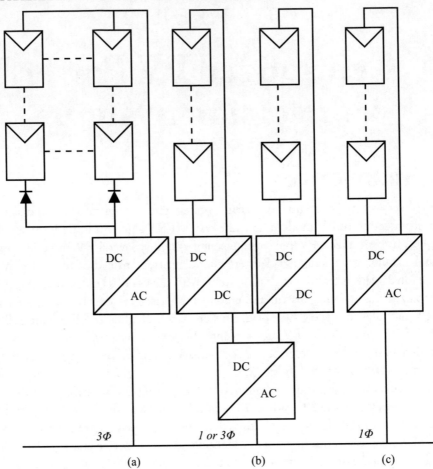

Figure 2.1: Centralized PV power conversion systems (a) central, (b) multi-string, and (c) string.

PV strings that have different orientations to a common power converter. But the drawback of this configuration is the need of a power conversion stages for each string which introduces power losses. Generally, central inverters are of three-phase type while multi-string-based systems can be either single- or three-phase.

In the central and string configuration, strings are directly connected to the DC-side of the central inverter and therefore the output voltage of each PV string is designed to meet the input voltage requirement of the inverter. Therefore, the output voltage of PV strings is large and high voltage DC cables are required to connect PV panels to the central inverter. As a result, maintenance and expansions are difficult in centralized PV systems. This limitation is relaxed to a certain extent in multi-string systems as there are separate DC-DC converters for each string that can be used to boost the input voltage in addition to tracking individual MPPs. Another significant problem common to all these centralized systems is shading of the PV modules in strings as string performance is completely degraded by the shaded modules.

2.2   CENTRAL INVERTER-BASED PV POWER CONVERSION SYSTEMS

Central inverters are utilized in large utility scale PV power conversion systems ranging from a few hundreds of kilowatts to a few megawatts. As mentioned above, in such systems parallel connected PV strings comprising series connected PV modules are interfaced to the utility grid using a central inverter. The number of PV panels in a single PV string and the number of parallel connected PV strings are decided by the input voltage requirement of the central inverter and the required output power of the entire PV system, respectively. PV strings are parallel connected in the generator junction box through string diodes, isolators, and fuses to block reverse current and to isolate strings when needed. The central inverter integrates PV array to the three-phase utility grid while performing global MPPT function for the entire PV system. Furthermore, the central inverter control system is responsible for providing other complex control tasks such as islanding detection when there is an outage of the utility grid with system condition monitoring and communicating with remaining parts of the system.

The complexity of DC wiring in the central inverter-based PV systems is reduced compared to string and multi-string inverter-based systems as it has comparably reduced parts count such as semiconductor devices and fuses. The scalability of the central inverter-based system is restricted due to the lack of modularity as compared to string-based systems.

Power converter topologies that are applied in central inverters should not have complex configuration in order to minimize power losses and to increase power density. Furthermore, the output current quality needs to be maintained with low harmonic distortion and ideally zero DC current content in order to comply with the country-specific grid codes and avoid saturation and heating of transformers in the grid. Stepping up of the power converter output voltage to grid voltage levels is another requirement if the output voltage is low. These basic requirements can be met with the use of suitable power converter topologies, filters, and transformers. Transformers provide required galvanic isolation and low electromagnetic interferences to the central inverter with the expense of increased power losses, weight and cost. With the development of semiconductor technologies, voltage and power ratings of switching devices are increased. As a result, transformerless grid connection of PV power converters has become a reality. These topologies possess high power density and efficiency with low weight and cost. However, such topologies necessitate high voltage at the output of strings so that they can meet the input DC-voltage requirements of the inverter. The central inverter topology can be of bridge type or multi-level type depending on the DC-bus voltage of the converter. The topology selection of grid inverter is based on several factors such as applied power modulation method, grid filter design complexity, and the cost.

An array of PV panels can be integrated into the utility grid using isolated single-phase power converter, as shown in Fig. 2.2a. In this configuration, the MPP of the PV array is tracked using the front-end DC-DC converter over a wide range of input voltage. The utility connected full-bridge converter transfers power to the grid. The control objective of the grid interfacing converter is to regulate the DC-link voltage so that the quality of the output current is maintained. When there is an increase in the captured PV power, the DC-link voltage tends to increase and the grid interfacing converter tries to regulate it by injecting more power to the grid resulting in an increase of the output power. The opposite happens when the captured PV power is reduced.

The front-end converter of the inverter shown in Fig. 2.2a can be replaced with a non-isolated boost converter as shown in Fig. 2.2b. The front-end boost converter not only tracks MPP of the PV array in this configuration but also provides required voltage boost to meet the DC-side voltage requirement of the inverter. However, in this configuration a care should be taken to use an appropriate mechanism to reduce leakage ground currents of the inverter resulting from parasitic capacitances of PV array due the absence of isolation transformer. Both topologies shown in Fig. 2.2 require electrolytic capacitors to mitigate double line frequency power ripple present in single-phase systems.

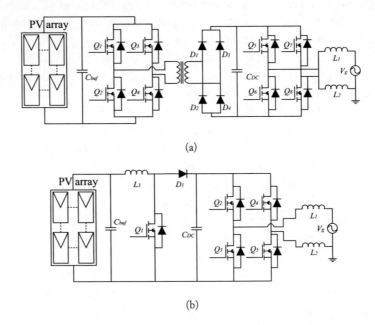

(a)

(b)

Figure 2.2: Central inverter-based PV power conversion systems with (a) isolated DC-DC converter, and (b) non-isolated DC-DC converter.

These single-phase connections can be replaced with three-phase connections in order to increase the output power of the inverter. Depending on the type of connection required, the power converter of the three-phase system can either be three-phase three-wire (3P3W) or three-phase four-wire system (3P4W). Recently, transformerless solutions gained more attention in three phase systems owing to their high-power processing capability with reduced losses, size, and cost.

The front-end converters of the two topologies shown in Fig. 2.2 are boost type converters while the grid-side converter is a buck type converter. This two-stage architecture essentially increases the component count which can be overcome using an interfacing converter that has both buck and boost capabilities. One such topology, known as Z-source converter, is shown in Fig. 2.3 [13]. The boost operation of this topology is obtained by introducing a shoot-through stage in power switch modulation scheme so that both switches in one phase leg are turned on simultaneously compared to a voltage source inverter where a short circuit would occur with such a gating pattern. This topology can be further modified using LF isolation transformer connected at the output of DC-AC converter in order to satisfy grid codes.

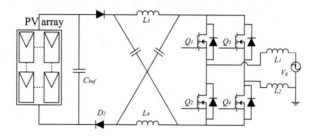

Figure 2.3: Central inverter-based PV system with an impedance source.

2.3   STRING-BASED PV POWER CONVERSION SYSTEMS

The underperforming PV arrays of central inverter-based PV power conversion systems can be partitioned using power converters connected to each PV string. A multi-string PV inverter integrates several such PV strings to the utility grid through a DC-DC converter connected to each PV string. This type of systems can be built using a single converter box with several embedded DC-DC converters and common operational control unit (OCU). The OCU starts/stops operation, maintains safety functions, conducts supervisory control of common inverter and communications. Such a multi-string-based system can be used to integrate PV strings with different voltage and current ratings, orientations and shading patterns. Hence, multi-string-based systems combine advantages of both string (partially distributed MPP) and central inverter-based systems (reduced number of inverters).

The number of parallel connected DC-DC converters depends on the overall output power of the multi-string inverter. The selection criteria of the grid-connected DC-AC inverter is completely based on the DC-DC converter topology and grid codes. The load isolation should be provided by the DC-AC inverter if it is a mandatory requirement and not provided by the DC-DC converter. In that case, the complexity of the DC-AC inverter is reduced as leakage ground currents are no longer a challenge in designing remaining parts of the inverter. Otherwise, there needs to be a proper mechanism to mitigate leakage ground currents using modifications in the topology, power modulation, and grid filter design as explained in later sections of this chapter. A non-isolated topology can be used as a DC-AC inverter if DC-DC converter provides isolation. In such a situation, design and topology selection of the remaining parts of the multi-string inverter is straightforward.

The DC-DC converter should possess voltage boost (/buck) capability and minimum device count to reduce losses. The required voltage amplification can be obtained using either active boost (/buck) solutions or a passive solution such as an isolation transformer. Boost and buck-boost converter topologies can be used as non-isolated options. The full-bridge, half-bridge, and push-pull converter topologies can be used as isolated single-ended solutions with reduced efficiency in high power applications due to their asymmetrical transformer core utilization. The topology selection should be done with extra care based on tracking capability of different converter topologies and their converter properties. Moreover, the resistance range of power converter topology and the load should also be taken into account in order to achieve proper MPPT. The diagram in Fig. 2.4 illustrates resistance range of buck and boost converters against the equivalent resistance of the PV array at the MPP. The requirement for maximum power transfer is that the input resistance of the converter should be equal to the equivalent resistance of the PV array at MPP.

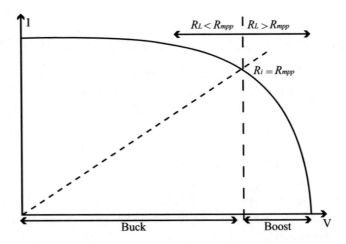

Figure 2.4: Possible ranges of load resistance for proper MPP tracking with buck type and boost type converters (for example, if $R_L > R_{mpp}$ a buck converter cannot be used to track the MPP).

The converter input impedance depends on the duty ratio of the converter which can be operated either in continuous conduction mode (CCM) or discontinuous conduction mode (DCM). For example, the input resistance (R_i) of the buck converter ranges from load resistance (R_L) to infinity with duty ratio varied from 0 to 1. In that case, load resistance (R_L) should be less than the equivalent resistance at the MPP (R_{mpp}) of the PV array to have proper MPP tracking. On the other hand, if the front-end converter of the PV array is based on a boost converter, the input resistance of the converter ranges from 0 to load resistance with the duty ratio varied from 0 to 1. In that case, load resistance (R_L) should be greater than the equivalent resistance at the MPP (R_{mpp}) of the PV array to have proper MPP tracking. These possible capture zones in terms of load resistance and operating ranges of buck and boost converters are illustrated in Fig. 2.4.

Multiple DC-DC converters can integrate several PV strings to the DC-bus of multistring inverter, as shown in Fig. 2.5. The front-end converter can be based on either non-isolated topology as shown in Fig. 2.5a or isolated topology, as shown in Fig. 2.5b. The grid-connected inverter based on either full-bridge or half-bridge converter is interfaced with the DC-bus to obtain a grid synchronized output current. Furthermore, multi-level inverter topologies can be used as grid connected inverter to minimize voltage and current ratings of the power semiconductor devices and thereby reduce cost and losses compared to bridge topologies as explained in Section 2.5. The multi-string PV inverters shown in Fig. 2.5 can be designed for single-phase operation. The power rating of the inverter can be increased using a three-phase system as explained Section 2.2.

2.4   GRID-CONNECTED INVERTERS

The DC-AC inverter plays a key role in grid-connected PV power conversion systems, irrespective of the PV module configuration. As mentioned above, grid-connecting inverters can be categorized as transformerless or grid-isolated based on the presence of a transformer between the grid and the inverter. Either high-frequency (HF) or low-frequency (LF) operation can be performed depending on the position of the isolation transformer within the inverter. The load isolation can be obtained by placing a transformer in the HF stage of the inverters, as shown in Figs. 2.2a and 2.5a as HF transformers have reduced weight, volume, and cost compared to the LF counterpart. The bridge converter topologies such as full-bridge and half-bridge can be used as shown in Figs. 2.2a, 2.3, and 2.5. The full-bridge and half-bridge converters are commonly used as grid-connected inverter due to their simple configuration, reduced number of semiconductor devices, and simple power modulation method. The grid-integration of the inverter can be realized using such a simple solution with the help of the grid isolation provided in the HF stage in order to fulfill grid codes requirements. Otherwise, LF transformers have to be connected when there is no isolation in DC-DC converter at the HF stage, as shown in Fig. 2.2b. But LF transformers need to be omitted from the inverters in utility scale applications rated up to several MW due to safety hazards in ground fault conditions. But this implementation causes DC currents to be injected into the utility grid and it needs to be overcome using improved switching control strategies and

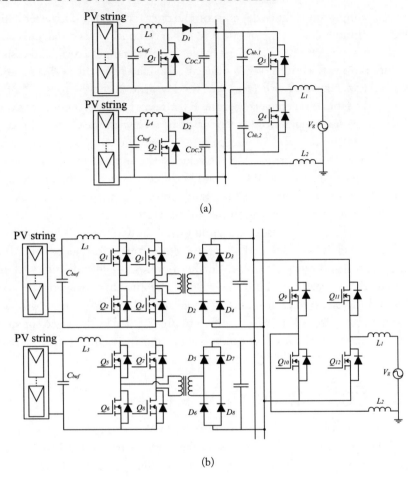

Figure 2.5: Multi-string inverters with (a) non-isolated, and (b) isolated DC-DC converters.

passive devices. Therefore, transformerless topologies can become a viable solution in integrating PV arrays comprising module strings.

However, transformerless systems result in leakage ground currents and common mode voltages due to parasitic capacitance between PV cells and panel structure as it becomes the only barrier to the common mode currents. The stray capacitances are due to planar structure of the PV cells and module glass. Moreover, the stray capacitance depends on the ground nature, weather and dust as well. The common mode currents conduct through stray capacitances which are in the order of hundreds of pico Farads and through transformers of grid isolated topologies. The impedance created by the stray capacitances can attenuate common mode current caused by low and medium order harmonics of the common mode voltage. In the case of transformerless topologies the common mode current cannot be mitigated using simple solutions.

An expression for the common mode voltage of the converter based on the bridge topology can be derived as in (2.1) with reference to Fig. 2.6a. The expression in (2.1) can be reduced to

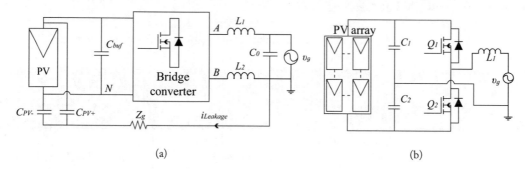

(a)   (b)

Figure 2.6: (a) Basic elements of the transformerless PV inverter and (b) its implementation with a half-bridge converter.

its simplest form as in (2.2) with similar or closer inductor values of L_1 and L_2. The common mode voltage and leakage ground currents are a function of output harmonic filter inductors. The leakage ground current can be eliminated by keeping the common mode voltage at a constant value. The half-bridge converter can be used to keep the common mode voltage at a constant value by connecting the neutral point of the utility grid to midpoint of the DC-bus capacitors, as shown in Fig. 2.6b. In that case, L_2 becomes zero and the expression in (2.1) can be further simplified to (2.2). The midpoint voltage should be kept at a constant value by giving a constant reference to the grid current controller. But the limitation of this configuration is evident as the voltage gain of a half-bridge converter is half of the gain of a full-bridge converter. Therefore, the input voltage of the half-bridge converter should be two times large compared to that of the full-bridge converter. In order to meet this requirement, there should be either a higher number of PV panels in a single string or a front-end step-up DC-DC converter. Therefore, in summary, neither the full-bridge converter nor the half-bridge converter in their simplest form is capable of eliminating common mode current and voltages without the isolation transformer.

$$v_{cm} = \frac{v_{AN} + v_{BN}}{2} + (v_{AN} - Vv_{BN})\frac{L_2 - L_1}{2(L_2 + L_1)} \tag{2.1}$$

$$v_{cm} = \frac{v_{AN} + v_{BN}}{2} + \frac{v_{AN} - v_{BN}}{2} = v_{AN} \tag{2.2}$$

$$v_{cm} = \frac{v_{AN} + v_{BN}}{2}. \tag{2.3}$$

There are many proposed solutions in mitigating leakage ground currents in full-bridge converter topologies ranging from modifications to converter topology or power modulation schemes to the installation of improved passive common mode filters. This can be further analyzed by considering full-bridge converter and its power modulation methods, namely unipolar, bipolar, and hybrid. Each power modulation scheme has its own advantages and disadvantages as summarized in Table 2.1 and it shows that they are not effective in achieving highest efficiency, low common mode current, and EMI at the same time.

Table 2.1: Comparison of common power modulation strategies of full-bridge converter

Power modulation strategy	Advantages	Disadvantages	Remarks
Bipolar	Low leakage current and EMI, higher filter requirement	Low efficiency due to high magnetic losses and reactive power	Small EMI filter size
Unipolar	High leakage current and EMI	High efficiency with zero voltage states	Suitable to be used with LF transformer, EMI filter is large without transformer
Hybrid	High leakage current and EMI, higher filter requirement	High efficiency due to lower reactive power exchange and lower switching frequency in one leg	Lower losses in power switches

The full-bridge converter in its simplest form is not suitable to be used as the inverter in transformerless grid connecting systems irrespective of the power modulation scheme used. Therefore, the basic full-bridge inverter can be modified, as shown in Fig. 2.7, to achieve high efficiency and lower common mode voltages and leakage ground currents. Figure 2.7a shows modified full-bridge converter with an additional active power switch and is known as H5 inverter. The additional switch in the positive link disconnects the input PV array from the utility grid at zero voltage state of the full-bridge inverter. There is high efficiency in the PV inverter due to lower magnetic losses in filter inductors, lack of reactive power exchange, and low leakage current and EMI. Most importantly, the input energy is not fed back into the DC-link capacitor. This inverter is used by SMA in their SunnyBoy inverter series with hybrid modulation to achieve high efficiency [14]. However, this inverter gives rise to a high leakage current when there are switching delays in the power switches. A modified PV inverter known as full-bridge with DC

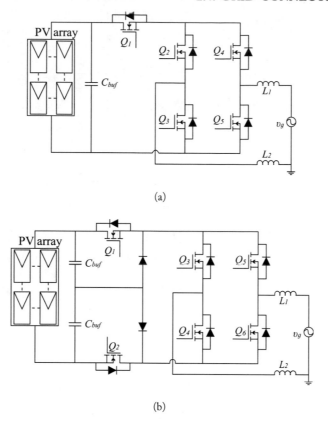

(a)

(b)

Figure 2.7: The modified PV inverter topologies based on full-bridge converter (a) H5 and (b) FB-DCBP. *(Continues.)*

by pass (FB-DCBP) is shown in Fig. 2.7b with DC bypass circuit comprising two switches in the DC-link and two clamping diodes [15]. The DC-link switches are turned-off at zero voltage states of the full-bridge inverter and diodes clamp voltages across switches and inverter output terminals to the midpoint of DC bus capacitors to obtain a constant common mode voltage. A PV inverter known as high efficient and reliable inverter concept or HERIC which is based on the full-bridge converter is shown in Fig. 2.7c with two AC decoupling switches [16]. The output of the inverter is short circuited using turn-on of the additional switches at the output of the inverter. The bipolar modulation is used with this inverter topology to obtain a high efficiency.

A modified PV inverter known as full-bridge zero voltage rectifier (FB-ZVR) is shown in Fig. 2.7d. The operation of this inverter is similar to the HERIC and it is based on AC decoupling technique at the zero voltage states of the full-bridge inverter. The clamping diode connected at the midpoint of the DC-bus helps keep the common mode voltage at a constant value and it also

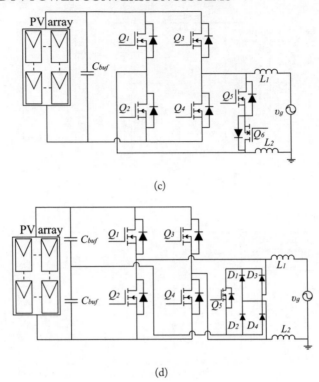

(c)

(d)

Figure 2.7: *(Continued.)* The modified PV inverter topologies based on full-bridge converter (c) HERIC and (d) FB-ZVR.

prevents discharge of DC-bus capacitors. This topology has high efficiency due to lower leakage current and EMI. Moreover, this topology has reduced magnetic losses in the output filter inductors as a result of low reactive power exchange between the grid and the inverter. Furthermore, there are topologies known as H6 inverters which mitigate leakage ground currents by providing alternative paths to AC and DC currents when necessary. The operation of all aforementioned PV inverters is based on the decoupling of AC and DC sides with the aim of mitigating leakage currents.

Moreover, leakage ground currents of inverters can be reduced using a connection between neutral of the grid connection and negative terminal of the PV array, as shown in Fig. 2.8a. The PV inverter shown in Fig. 2.8a is derived using buck-boost topology and it is known as Karschny inverter [16]. A PV inverter based on a similar strategy is shown in Fig. 2.8b and it is a derivation of the buck converter. These two inverter topologies are highly reliable even though they are not capable of providing reactive power. A PV inverter topology based on virtual DC bus is proposed to overcome leakage current by connecting grid neutral to the negative terminal

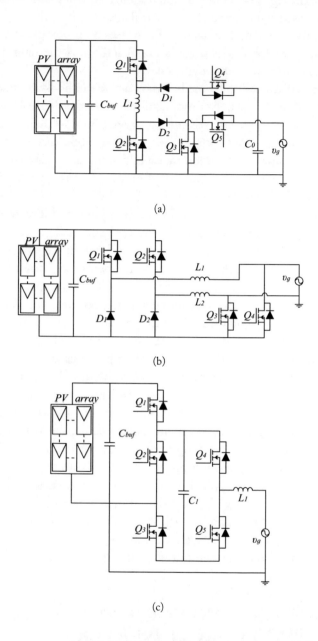

(a)

(b)

(c)

Figure 2.8: PV inverters based on (a) buck-boost (Karschny), (b) buck, and (c) virtual DC-bus.

of PV array, as shown in Fig. 2.8c. Unipolar and double frequency sinusoidal PWM methods can be used with the proposed topology to eliminate leakage currents. This topology is more suitable for low power applications and additional current stresses on power devices and capacitors can be limited. However, all these topologies use additional active switches and passive components to reduce leakage currents and obtain high efficiency.

All the solutions discussed so far are based on active methods and they require extra power switches and modifications to the power modulations methods. But leakage ground currents can be reduced using relatively simple and reliable passive methods as well. Considering the fact that in the absence of an isolation transformer, the common mode currents flown into the ground are limited only by the converter common mode impedances, an EMI filter can be designed to attenuate common mode voltages with the aim of preventing strong ground currents. A few such passive common mode filter designs are shown in Fig. 2.9.

Common mode current can be reduced effectively using a LCL common mode filter which has a capacitor connected to the ground as shown in Fig. 2.9a. The ground connected common mode capacitor (C_{cm}) provides a low impedance path for high frequency current. The low frequency and DC currents see a high impedance which helps to reduce common mode currents. The effectiveness of mitigation of leakage currents can be further improved using a cascaded filter structure as shown in Fig. 2.9b. This cascaded filter is based on a LC common mode filter and a LCL harmonic filter. The common mode currents circulate through this filter instead of the DC power source of the system. A simple common mode filter shown in Fig. 2.9c has a reduced number of magnetic components. This filter structure provides an output current with low harmonic distortions and reactive power capability with the use of two inductors at the output of the full-bridge converter acting as part of the differential and common mode filters to minimize leakage current. Ground resistance (R_g) acts as a damping resistor while two resistors (R_{cm}) in the common mode filter are designed to prevent unwanted oscillations with the loss of small amount of power in the resistors. Proper filter design is a mandatory requirement in grid-connected PV inverters as there could be strong ground current with the saturation of filter components.

Similar approaches can be applied for three-phase systems as well in order to minimize common mode voltages that induce large leakage currents in non-isolated PV inverters. However, filter components such as inductors and capacitors need to be designed with better insight in order to reduce volume, cost, and other adverse effects that may arise with three-phase operation. Power modulation schemes can also be needed to reduce the strength of common mode voltages to an acceptable level.

2.5   MULTI-LEVEL CONVERTER TOPOLOGIES FOR GRID-CONNECTING INVERTERS

Multi-level inverters are proposed to be used in PV applications in order to improve the efficiency, reduce harmonic distortions in the output current, reduce device ratings and thereby reduce cost, reduce power losses, reduce leakage current due to constant common mode voltages,

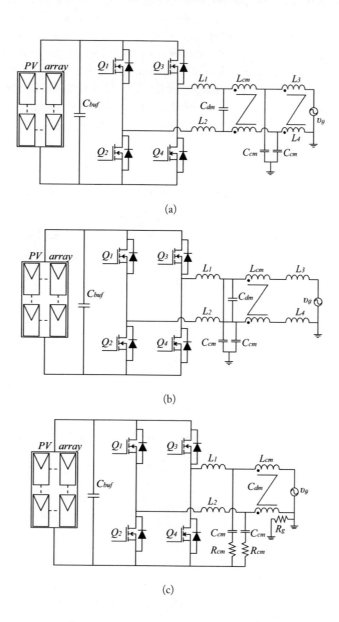

Figure 2.9: Passive common mode filter designs for transformerless inverters with common mode capacitor connected to the (a) ground, (b) DC-bus, and (c) harmonic filter.

and to lessen complexity of the output filter requirements. There are numerous types of multi-level inverter topologies reported in literature. All these topologies can be divided into four categories: namely, diode-clamped topologies, flying capacitor topologies, cascaded topologies, and modular topologies. An example circuit for each category for the single phase implementation is shown in Fig. 2.10. The type and number of levels in the output of the multi-level inverter depends on several factors such as number of semiconductor devices, DC-side voltage levels of the converter modules, and the modulation scheme. Generally, voltage and current ratings of the semiconductor devices vary with the number of levels in the output voltage.

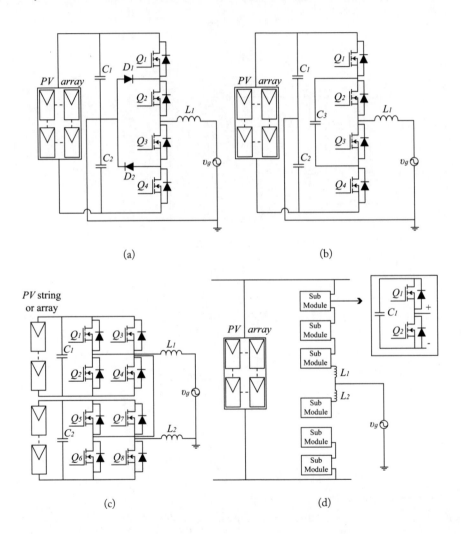

Figure 2.10: Multi-level inverter topologies (a) diode-clamped three-level inverter, (b) capacitor-clamped three-level inverter, (c) cascaded multi-level inverter, and (d) modular multi-level inverter.

The multi-level inverters shown in Figs. 2.10a, b, and c are more suitable for applications that require low output and DC-bus voltages. The multi-level inverter shown in Fig. 2.10d is more suitable for applications where there are medium voltage inputs and outputs.

2.5.1   DIODE-CLAMPED MULTI-LEVEL INVERTER (DCMLI)

The three-level diode-clamped multi-level (also known as neutral point calmed (NPC)) inverter can be used in single- and three-phase system. The schematic of a single-phase three-level diode-clamped inverter is shown in Fig. 2.10a. In order to derive the corresponding three-phase three-wire inverter, the switch-diode arrangement has to be repeated twice. If a multi-phase NPC inverter is required the circuit shown in Fig. 2.10a needs to be modified using required number of legs connected in parallel. The neutral point of the utility grid is connected to the midpoint of the front-end DC bus in either case.

When calculating common mode voltage, the Eq. (2.1) can be simplified into (2.2) due to the absence of L_2 in multi-level inverters and hence the leakage current due to the varied common mode voltages can be kept at a low value. In order to ensure low common mode voltages, the midpoint voltage of the DC-bus should be kept at a constant value, ideally at half of the DC-link voltage. There are number of ways reported in literature to achieve this. Among them, the proper selection of redundant switching states is the most common technique. With the use of these techniques, the DC-bus voltage can equally be shared among active power switches of the NPC inverter. But the blocking voltages of clamping diodes will not be equal especially in the inverters with more voltage levels in the output. Moreover, the power losses in the outer switches are different from that of the inner switches. These issues can be solved with the use of active switches in place of clamping diodes. The resultant topology is known as the active neutral point clamped (ANPC) inverter. Both NPC and ANPC exhibit lower switching and conduction losses compared to full-bridge derived topologies such as H5 and hence show higher efficiency. An active switch-based three-level NPC converter derived upon conventional two-level inverter is shown in Fig. 2.11 [17]. In this particular configuration there is a bidirectional active switch connected to the midpoint of the half-bridge converter. The active power switches of the half-bridge have two times higher device rating compared to the other two active switches. This configuration has reduced filter requirements due to the presence of three-level output and lower switching losses compared to the half-bridge converter. However, higher conduction losses are observed due to the increase of switching devices in the conduction path. The performance of the inverter can be further improved using a single device module in the middle arm with reverse voltage blocking capability, as explained in Section 2.6.

PWM-based power modulation schemes such as sinusoidal PWM (SPWM), space vector modulation (SVM), and selective harmonic elimination (SHE) can be used to modulate the inverter switches. SPWM method is based on comparison between two triangular carrier waveforms with a sinusoidal reference waveform. This method helps to reduce harmonic distortion of the output voltage or current. The operation of SVM is based on generation of switching se-

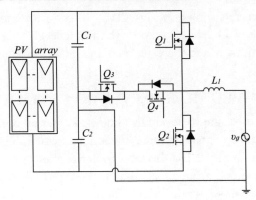

Figure 2.11: Three-level inverter based on active switches and half-bridge converter.

quences of the multi-level inverter to obtain a desired output voltage level with the balancing of voltages of DC-bus capacitors. SHE-based power modulation methods are applied to reduce semiconductor losses by maintaining required harmonic levels with reduced modulation indices. The power density of the NPC-based PV inverters can be improved using LCL harmonic filters as leakage current is not significant with the connection of neutral point of the output to the mid-point of the DC-link.

2.5.2   CAPACITOR-CLAMPED MULTI-LEVEL INVERTER (CCMLI)

A single-phase three-level capacitor-clamped inverter is shown in Fig. 2.10b which uses capacitors to clamp active switch voltages. The number of clamping levels can be increased with the addition of switches and clamping capacitors. The capacitor-clamped multi-level inverter is also known as flying capacitor inverter as the clamping capacitors are not directly connected to the positive or negative terminals of the supply. The number of active switching states is increased compared to NPC inverter as capacitors do not block the reverse voltage. The capacitor voltage is rated to $V_{DC}/(n-1)$ where n is the number of levels in the output voltage. The capacitor voltage balancing is possible in this topology as an identical current conducts through all active switches and redundant switching states can be used to further optimize this operation.

2.5.3   CASCADED MULTI-LEVEL INVERTER (CMLI)

Cascaded multi-level inverters (CMLIs) use a different approach to obtain multi-level output voltages compared to NPC and capacitor-clamped multi-level inverters. In CMLIs isolated DC sources are used at the DC-link of the full-bridges, also known as H-bridges, as shown in Fig. 3.10c. A single H-bridge in Fig. 2.10c can generate three voltage levels at the output as V_{dc}, 0 and $-V_{dc}$. When it is cascaded with another H-bridge, as shown in Fig. 3.10c five different voltage levels can be obtained at the output as $2V_{DC}$, V_{DC}, 0, $-V_{DC}$, and $-2V_{DC}$. Cascaded multi-level inverters can be used in single-phase applications as well as in three-phase applications. The number of power semiconductor devices required to obtain a certain number of levels in the output voltage is less in the CMLIs compared to NPC and capacitor clamped inverters, especially at higher levels (> 3). Moreover, the modular architecture makes expansion and replacement straightforward.

Series connected PV strings can be connected as DC sources of full-bridge converter modules, as shown in Fig. 3.10c. However, the output voltage of each PV string should be similar to obtain an output without significant distortions. In such a case each PV string can be interfaced to the full-bridge converter through a DC-DC converter in order to mitigate voltage variations caused by partial shading of PV array.

2.5.4 MODULAR MULTI-LEVEL INVERTER (MMLI)

The basic configuration of the modular multi-level inverter which is suitable for medium voltage applications as well is shown in Fig. 2.10d. Its building block, also referred to as sub module, can be a half-bridge converter as shown in Fig. 2.10d, a full bridge converter or a multilevel converter such as NPC. Capacitors can be used as the DC-source of the sub modules together with a suitable switching logic to maintain the capacitor voltage. In the single-phase implementation, the maximum output voltage of this inverter is equal to the DC-bus voltage and the desired stepped output can be obtained by adding, subtracting or bypassing the outputs of sub modules. The positive and negative output voltages are obtained using arms comprising of n series connected sub modules connected at a common point as shown in Fig. 2.10d [18]. There are two inductors at the common connecting point of the arm in each phase to cater voltage differences at the turn-on and off of the sub modules.

Modular multi-level converters are still in research stage for them to be effectively used in real PV applications with the need of necessary improvements in dynamic voltage balancing and losses with circulating currents. However, this configuration has a unique advantage of having a single source compared to a cascaded system. Several power modulation schemes are proposed to overcome above mentioned problems such as amplitude modulation methods, phase modulation methods and selective virtual loop mapping methods. The number of power semiconductor devices in this topology depends on the number of sub modules in a phase leg.

2.5.5 COMPARISON OF MULTI-LEVEL INVERTER TOPOLOGIES

Common multi-level inverter topologies that can be used in centralized PV systems are discussed above. In addition to those topologies, there are other multi-level inverter topologies such as hybrid and asymmetric hybrid types that are mainly based on existing multi-level inverter topologies and their combinations. In asymmetric hybrid inverters, DC supply voltages of the sub modules are not equal. Moreover, their switching frequencies can be different from each other. For example, in an asymmetric hybrid inverter, a sub module with a lower DC-supply voltage can be switched at a higher frequency compared to that of a sub module having a larger DC supply voltage. This reduces power losses within switching devices. Table 2.2 shows a comparison of single-phase multi-level converter topologies that can be used as a grid connecting inverter in centralized PV systems where n is the number of levels in the output voltage and V_{DC} is the DC-bus voltage of the multi-level converter.

Table 2.2: Topology comparison based on active and passive devices

Topology	No of Power switches	No of diodes or capacitors	Component voltage rating	Passive devices
DCMLI	*2(n-1)*	Diodes - (n-1)(n-2)	V_{DC}/ n-1	0
CCMLI	*2(n-1)*	Caps - (n-1)(n-2)/2	V_{DC}/ n-1	0
CMLI	*2(n-1)*	0	V_{DC}/ n-1	0
MMLI	*4(n-1)*	0	V_{DC}/ n-1	2

2.6   CONTROLLER DESIGN OF CENTRALIZED PV POWER CONVERSION SYSTEMS

The controller design of the central, string- and multi-string inverters entirely depends on the interconnection between power converters and topology of the each power converter. In the case of a multi-string system having separate DC-DC converters, the MPP of each string can be tracked individually with proper control of these DC-DC converters. The DC-AC inverter can be used to generate grid synchronized output current while maintaining the DC-bus voltage at a pre-defined value to guarantee a high quality output current. As explained in Sections 3.2 and 3.3 in string and central architecture-based centralized PV systems, there is only one grid-connecting inverter. The PV strings can be connected to the grid interfacing inverter directly or through another power processing stage and the two approaches are called single-stage and double-stage, respectively. The controller of the double stage central and string inverters exhibit a form similar to that of multi-string inverter as there are two power conversion stages. However, the controllers of string and central inverters take a different form if there is only a single power conversion stage as it needs to provide all the functionalities such as MPPT, grid current control and synchronization with grid voltage. The system characterization of a single-stage PV inverter is presented in the following section using a mathematical model that explains dynamic properties of the system.

2.6.1   MODELING OF SINGLE-STAGE CENTRAL INVERTER

The basic configuration of the transformerless single-stage central inverter based on a full bridge converter is shown in Fig. 2.12 with parasitic capacitances of the PV array. The PV array is integrated to the DC-link of a full-bridge converter and a grid filter comprising an inductor is used to filter out harmonics.

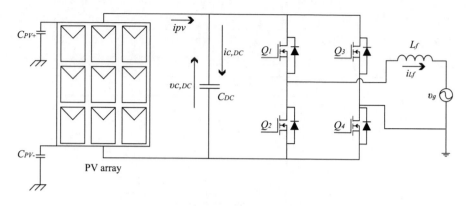

Figure 2.12: The basic configuration of a single-stage grid-connected transformerless central inverter.

Dynamic behavior of the inverter shown in Fig. 2.12 can be described using (2.4) and (2.5) that are obtained applying Kirchhoff's voltage and current rules when $Q_{1,4}$ (during positive half cycle of the grid voltage) and $Q_{2,3}$ (during negative half cycle of the grid voltage) conduct currents assuming $v_g = V \sin \omega t$:

$$C_{DC} \frac{dv_{C,DC}}{dt} = -Si_{L,f} + i_{PV} \qquad (2.4)$$

$$i_{L,f} \frac{di_{L,f}}{dt} = Sv_{C,DC} - v_g. \qquad (2.5)$$

The switching function S can be either -1 or 1 based on switching control signals applied to power switches $Q_{1,4}$ and $Q_{2,3}$ by the grid current controller of the inverter. The grid filter can be replaced by either a LC or LCL filter in order to obtain improved performance in terms of ripple attenuation, harmonic trapping and volume of the filter as explained in [19].

A 3.25 kW central inverter is simulated with an LCL filter and 25 kHz switching frequency using an unipolar power modulation method. Figure 2.13 shows the waveforms of grid voltage, inverter output current, and power. The 50 Hz grid voltage and output current create a 100 Hz power output as shown in Fig. 2.13. The 100 Hz power ripple propagates to the DC-bus and beyond if DC-bus capacitor is not able to mitigate it as shown in Fig. 2.14. Figure 2.14 shows DC-bus voltage, PV array output current and power when there is 3 mF DC-bus capacitor. A similar set of waveforms are shown in Fig. 2.15 when DC-bus capacitor is replaced with a 0.5 mF capacitor. There is an insignificant DC-bus voltage and current ripple with a 3 mF capacitor as compared to the case of 0.5 mF capacitor. It is necessary to have reduced voltage ripple as shown in Fig. 2.14 in order to track MPP of the PV array efficiently due to the fact that input voltage

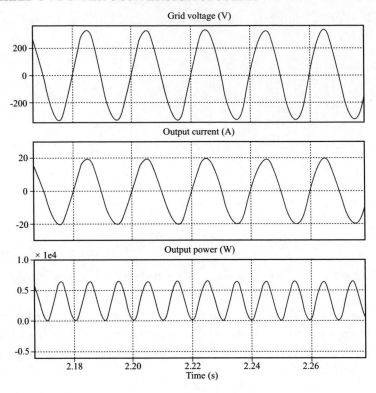

Figure 2.13: Simulation results showing the output of a central inverter, from top to bottom the grid voltage (V), inverter output current (A), and power (W).

control type MPPT algorithms have better stability compared to that of input current control methods as PV arrays behave like a current source.

Electrolytic capacitors are needed to provide this capacitance with reasonable volume and cost. But their performance may degrade with time due to electrical, thermal and mechanical stresses. The main sources of the degradation are high temperature rise and large current ripples which usually occurs in PV inverter operation. The degradation gives rise to reduction in capacitance and increase in equivalent series resistance (ESR) of the capacitor bank which could create instabilities within the system. Hence, proper monitoring mechanisms are required to estimate remaining life time of the capacitors in order to prevent possible catastrophic failures. The remaining life time of capacitors can be estimated using real capacitance and ESR at the measuring instant and ageing algorithms. The measurements that are required to calculate ESR and capacitance can be obtained using governing equations of front-end DC-DC converters.

Furthermore, leakage currents flowing through DC components of the system such as PV array have a significant effect on the reliability of the system. The leakage current is induced by

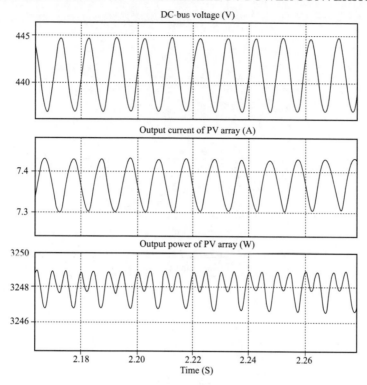

Figure 2.14: Simulation results showing the ripples in the DC side with 3 mF DC capacitance, from top to bottom the DC-bus voltage (V), PV array output current (A), and power (W).

common mode voltages. Figure 2.16a shows the common mode voltage at the negative DC rail of the PV array, obtained by assuming a 400 nF parasitic capacitance between the ground and negative rail. The zoomed waveform is shown in Fig. 2.16b which clearly shows negative DC rail voltage comprising components from common mode voltage and 100 Hz ripple.

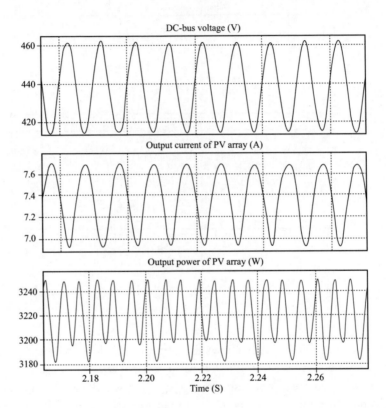

Figure 2.15: Simulation results showing the ripples in the DC side with 0.5 mF DC capacitance, from top to bottom the DC-bus voltage (V), PV array output current (A), and power (W).

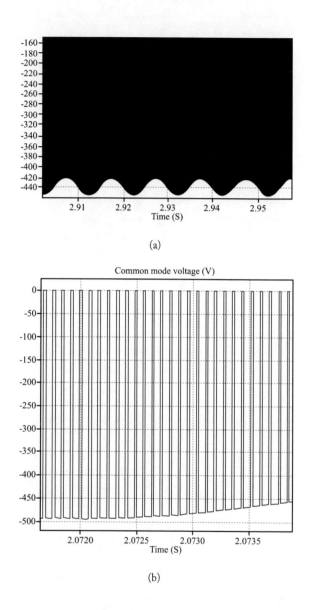

(a)

(b)

Figure 2.16: Common mode voltage (V) of negative DC rail of the PV array.

CHAPTER 3

Distributed PV Power Conversion Systems

3.1 INTRODUCTION

Partial shading, module mismatch, uneven aging, and soiling are the most common challenges that have to be taken into account when optimizing PV energy conversion efficiency, particularly in string and multi-string PV power conversion systems. Moreover, due to uneven temperature distribution across large PV arrays, individual modules may operate at different temperatures resulting in differences in their characteristic curves and MPPs. As mentioned in Section 2.1, if a centralized power conversion system is used to track the global MPP, module level MPPs may not align with the global MPP due to these differences in individual PV modules. The distributed PV power conversion systems comprising module level power converters are proposed to overcome this drawback.

The basic concept of the distributed systems is similar to that of the string and multi-string PV power conversion systems. In string-based systems, each string of PV modules is connected to the central inverter via a DC-DC converter and therefore string level MPP can be tracked through this converter. This concept has been further extended into the module level in distributed systems to overcome module mismatch, uneven aging, partial shading, and uneven temperature distribution. But, the number of power converters in fully distributed systems increases due to their per module architecture. Hence, distributed architectures are not yet feasible to be used in large utility scale PV power conversion systems rated to several hundreds of kilowatts and megawatts. However, there are reduced constraints in distributed systems in scaling up their power due to plug and generation capability of the building modules. Furthermore, there is reduced cost per unit power (cost per watt) generated using distributed PV systems which is one of the major requirements in small and medium scale residential applications. The overall cost of PV power conversion systems is the sum of costs of PV panels, balance of the system, installation and maintenance. The maintenance and installation cost contributes to higher portion of total systems cost due to much needed requirement of highly skilled labor force to work with high voltage DC power.

Each PV module of fully distributed PV power conversion systems which are used in domestic and small scale applications such as building integrated PV systems (BIPV) consists of its own power converter. The module integrated power converter of these PV power conversion systems are known as micro inverters and micro converters depending on the nature of their output. The micro inverter is a DC-AC converter integrated into back side of a PV panel and the whole unit is known as an AC module. An AC module integrates a single PV panel to the utility grid without intermediate power conversion stages. The micro converter, also known as a DC optimizer is a DC-DC converter with a high voltage-conversion ratio. A micro converter integrates single PV panel into a common DC-bus rated from 200–400 V. Hence, they should possess high voltage-conversion ratios to boost low PV module output voltage (36–72 V) to a high DC-bus voltage. The DC-bus is connected to a central grid interfacing inverter. As a result, this configuration can give rise to a single point of failure of the system which does not exist in micro inverter-based distributed PV systems. However, DC-optimizers have a dedicated power conversion stage to track MPP of PV panels which can be used to cover a wide range of the input voltage. This advantage cannot be observed in single-stage micro inverter-based systems as the inverter has to perform both MPPT as well as output current control to satisfy grid code requirements. Moreover, users do not need to have high technical competency in integrating AC modules to the utility grid as they are made grid compatible and this advantage cannot be observed in micro converter based systems. However, in either case, PV module integrated converters need to possess high reliability, efficiency, power density, and low weight and cost. Hence, all module integrated converter topologies tend to operate in HF as opposed to that of the centralized systems. As a result soft-switching techniques such as zero-voltage-switching (ZVS) and zero-current-switching (ZCS) are frequently used in these power converters to reduce switching losses as they are increased in proportion to the switching frequency. Hence, the selection of converter topology and associated power modulation scheme plays a key role in improving efficiency and power density without additional active and passive elements such as snubbers to enhance switching characteristics and to reduce electro-magnetic interferences.

3.2   DISTRIBUTED PV SYSTEMS WITH MICRO INVERTERS

The basic configuration of a single phase micro inverter-based PV power conversion system is shown in Fig. 3.1a. PV modules are parallel connected to the utility grid using AC modules. The voltage and current ratings of the micro inverter should be compatible with the associated PV module and the grid. This basic configuration has been slightly modified using "duo micro inverter" introduced by Enecsys to integrate two parallel connected PV modules to the utility grid using single power converter aiming to minimize the number of power converters [20]. A similar strategy can be implemented using strings comprising a minimum number of PV modules such as two or three to mitigate the effects of partial shading and other inherited drawbacks.

Micro inverters should operate the PV module at its MPP while injecting sinusoidal output current to the utility grid with low distortion. Moreover, other ancillary functionalities such as islanding detection, condition monitoring and communication should also be performed by micro inverters. As shown in Fig. 3.1b, micro inverters can be categorized into different groups based on the grid isolation, availability of a DC-link and the nature of the DC-link. In this categorization, micro inverters are broadly divided into two groups as transformerless and grid-isolated

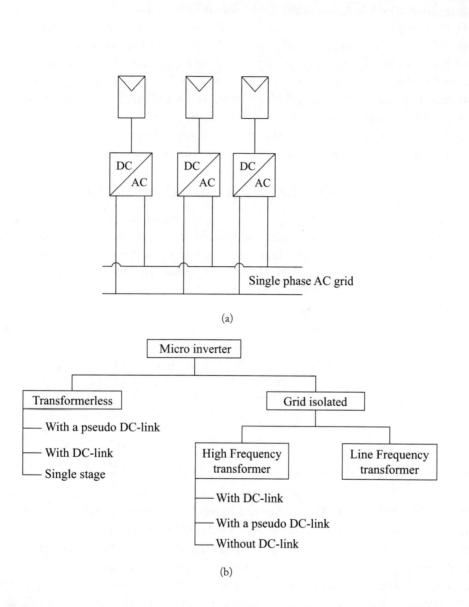

Figure 3.1: Micro inverter-based PV power conversion systems (a) basic configuration and (b) topology classification.

types considering load isolation as the main root. The transformerless micro inverters have highest efficiency and power density and can be used in countries where there is no tight regulation on load isolation and leakage ground currents. The load isolated micro inverters can be further divided based on the operating frequency of the isolation transformer. HF and LF transformers are utilized in micro inverters to provide a passive voltage boost to low input voltage and to minimize leakage ground currents. Both transformerless and grid isolated micro inverters can be further classified using the next dominant root of classification known as DC-link as shown in Fig. 3.1b. The DC-link plays a key role in micro inverters as it can be used to improve efficiency, cost, and reliability of the micro inverter as discussed in the following sections.

3.3   TRANSFORMERLESS MICRO INVERTERS

The copper and core losses of the magnetic components such as transformers and inductors are major sources of loss in power conversion systems. Moreover, transformer account for larger percentage of area, weight, and volume. Therefore, transformerless power converters show improved efficiency and increased power density. However, they require-alternative mechanisms to obtain the passive voltage gain provided by isolation transformers. As an alternative, an active voltage boost stage can be incorporated at the front-end of the transformerless micro inverters as shown in Figs. 3.2a and 3.2b. The front-end boost stage can be used to obtain an output that gives either a constant DC voltage or a rectified output that follows sinusoidal envelop compatible with the input of the grid connected DC-AC inverter. Hence, the operation of the grid connected stage can be either an inverter or unfolding circuit. The DC-link can also be omitted in order to reduce number of power conversion stages, to further improve power density and efficiency and to reduce cost. The basic configurations of such transformerless micro inverters are depicted in Fig. 3.2 as categorized in Fig. 3.1.

3.3.1   TRANSFORMERLESS MICRO INVERTER WITH A DC-LINK

The transformerless DC-link micro inverters consist of two power conversion stages are shown in Fig. 3.2a. The MPP of the PV module connected to the micro inverter is tracked by the front-end DC-DC converter. The grid connected DC-AC inverter generates low-frequency sinusoidal output currents and voltages. Two power conversion stages are interfaced through a DC-link with a sufficiently large DC capacitor in order to prevent the propagation of double line frequency power ripple into the front-end converter and the PV panel. However, the DC-link capacitor has been identified as the most dominant failure component and therefore more reliable capacitor types such as film capacitors are preferred as the DC-link capacitor. As discussed in Chapter 2, the DC-link capacitance can be further reduced by allowing the DC-link voltage to vary within a safe range.

The grid-connected DC-AC inverter transfers power to the load using either full-bridge, half-bridge, or multi-level inverter topologies as shown in Fig. 3.3. The three level NPC inverter as shown in Fig. 3.3c is a feasible option when compared its quality of the output current, number of

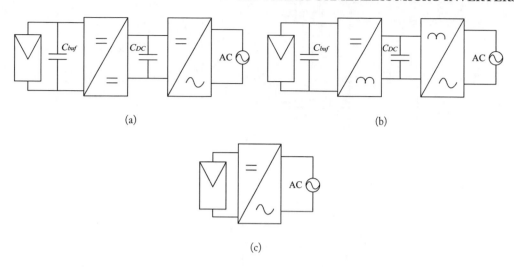

Figure 3.2: The basic configurations of the transformerless micro inverters with (a) DC-link, (b) pseudo DC-link, and (c) single power conversion stage.

semiconductor devices, and output filter design with other multi-level inverter topologies having higher number of levels [21]. However, proper power modulation methods need to be used to balance the voltage of the DC-link capacitors and to maintain leakage ground currents to a desired level. The requirement of complex power modulation strategies to reduce ground current can be relaxed to a certain extent using half-bridge converter, as shown in Fig. 3.3b, due to the connection between neutral point of the load and middle point of the DC-link. However, the voltage in the DC-link of the micro inverter with half-bridge converter is two-times higher compared to that of the micro inverter with the full-bridge converter as shown in Fig. 3.3a. Hence, there should be a compromise between the number of power switches and the required gain of the front-end converter when selecting a DC-AC inverter topology. The majority of transformerless DC-link micro inverters are based on these configurations.

The front-end DC-DC converter should have voltage boosting capability in order to provide active voltage boost to low PV generated voltage. Hence, front-end converter can be either boost converter, as shown in Fig. 3.3, buck-boost or other boost type of DC-DC converter. The selection criteria of the DC-DC converter can be based on power loss, boost ratio and characteristics of the input current. The continuous input current is a mandatory requirement to maintain an efficient PV power conversion in module level. Switching and conduction losses of the power semiconductor devices need to be reduced to minimize losses. The reduction of number of power switches gives rise to reduced conduction losses and it can be further optimized using interleaved topologies. Furthermore, a resonant operation can be implemented using leakage inductances and output capacitances of power switches to turn-on and off them with ZVS and ZCS without com-

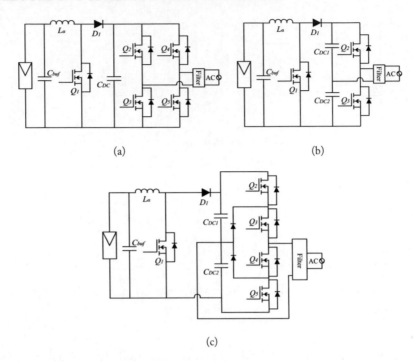

Figure 3.3: Transformerless micro inverters with grid-connected (a) full-bridge, (b) half-bridge, and (c) three-level inverter.

plex power modulation strategies and topology modifications. A few such transformerless micro inverter topologies that use aforementioned strategies are shown in Fig. 3.4.

Front-end converter of the micro inverter shown in Fig. 3.4(a) has a parallel input and a series output to boost PV module input voltage to a desired level [22]. The power losses of the front-end converter are reduced by processing a portion of power generated by the PV module to minimize conduction and switching losses. The front-end converter of the micro inverter shown in Fig. 3.4(b) follows a similar strategy to obtain a high boost ratio by having a series output of boost and buck-boost converters [23]. Moreover, in this topology the common ground line between the PV panel and the utility grid mitigates the leakage ground currents of the micro inverter. Coupled inductors can also be used to obtain a high boost ratio depending on the turns ratio between the inductors and the duty ratio of the power switch control signals. A micro inverter with a coupled inductor and a tapped winding based front-end converter is shown in Fig. 3.4(c) [24]. Power density of the micro inverter can be improved with coupled inductors instead of using separate magnetic components but at the expense of a complex configuration due to the tapped inductor.

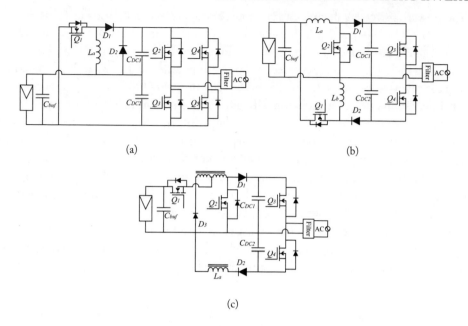

Figure 3.4: The transformerless micro inverter with front-end DC-DC converters with (a) parallel inputs and series output, (b) boost and buck-boost, and (c) coupled inductor.

3.3.2   TRANSFORMERLESS MICRO INVERTER WITH PSEUDO DC-LINK

The transformerless micro inverter with a pseudo DC-link also consists of two power conversion stages, as shown in Fig. 3.2b. In this topology, the passive power decoupling of the micro inverter is shifted towards the input of the front-end converter as there is a pseudo DC-link between the front-end converter and the grid connecting inverter. As a result, sufficiently large electrolytic capacitors are needed at the input of the front-end converter in order to prevent the interaction of double line frequency power ripple with the operation of the MPP tracker. The front-end DC-DC converter generates a rectified current that follows an envelope of the 50 Hz sinusoidal current operating in the discontinuous conduction mode (DCM). The grid-connected DC-AC inverter unfolds the rectified input to obtain LF output currents and voltages by switching active power switches at a low switching frequency. The DC-AC inverter can be based on either power MOSFETs, as shown in Fig. 3.5, or on thyristors or IGBTs with low turn-on resistance to reduce conduction losses due to the fact that conduction losses are more dominant when the switching is done at a lower frequency. Moreover, micro inverters with fully controlled power switches (MOSFET and IGBT) always show a better performance compared to a naturally commutated device- (thyristors) based solution when considering the harmonic content of the output current.

The power converters having current source properties can be used as front-end converter of the pseudo DC-link-based micro inverters. Two micro inverters with front-end converters that exhibit buck-boost operation are shown in Figs. 3.5a and 3.5b. The PV module of the micro inverter shown in Fig. 3.5a is integrated in opposite polarity as the output of the conventional buck-boost is reversed compared to buck and boost converters [25]. A micro inverter with front-end buck-boost converter that gives a positive output is shown in Fig. 3.5b and it has a higher number of power semiconductor devices [26]. The front-end converter of the micro inverter can be operated in either buck or boost mode depending on the control of the two active power switches. But mode change of the front-end converter should be controlled using a proper algorithm in order to minimize transients at the mode transition. A similar front-end converter characteristic can be obtained using switched inductor topologies with higher converter gains, as shown in Fig. 3.5c [27]. Moreover, micro inverter configurations with front-end DC-DC converters that are based on interleaved boost cascaded with buck converter topologies can be found in literature [28]. The interleaved boost cascaded with the buck converter shows a higher efficiency compared to buck-boost-based converter topologies [28].

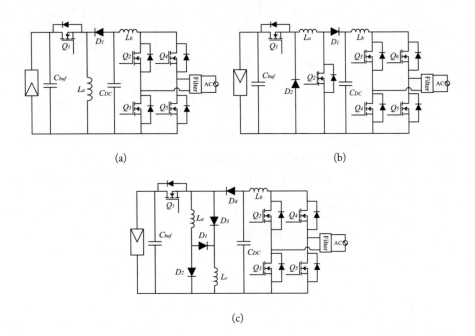

Figure 3.5: Transformerless micro inverters with pseudo DC-link and (a) buck-boost, (b) two switch buck-boost, and (c) switched inductors.

3.3.3   SINGLE-STAGE TRANSFORMERLESS MICRO INVERTERS

As the name implies, there is only one power conversion stage in these micro inverters instead of the front-end DC-DC converter and the grid-connected DC-AC converter found in two stage systems. Therefore, the micro inverter has to provide all the required functionalities such as MPPT, voltage boosting, injecting low distortion sinusoidal current to the utility grid and other ancillary services. Due to the absence of an intermediate converter, single-stage micro inverters have the highest power density and efficiency with the minimum number of power semiconductor devices and reduced power losses compared to two-stage power converters discussed above. However, reliability of these micro inverters get affected by the need for a large electrolytic capacitor at the input in order to prevent the propagation of the double line frequency power ripple to the PV module. Moreover, such topologies should have proper grid filters and switching control strategies to reduce leakage ground currents and common mode voltages that may arise due to parasitic capacitances of PV modules.

The functionalities provided by the DC-DC converter in two stage systems can be integrated into the single-stage micro inverter. Two such topologies are shown in Fig. 3.6. In order to realize the DC-DC converter operation in these topologies, power switches are required to operate at high switching frequencies. In simple boost converter integrated micro inverters, distortions may occur in the output current zero crossing as the boost converters are not able to generate an output voltage lower than their input voltage. This drawback can be overcome using the micro inverter shown in Fig. 3.6a which is based on differential mode operation. In this topology, two DC-DC converters based on either boost or buck-boost operations are controlled with 180° phase difference to generate an unipolar sinusoidal output. Alternatively, hybrid solutions with topological and power modulation modifications can also be used to overcome the aforementioned limitations. For example, tri-state modulation can be used with an additional power switch in order to freewheel inductor current in the boost converter at the zero crossing of the micro inverter output current [29].

A Z-source configuration used in centralized inverter systems can be adapted to a micro inverter for increased converter gain. The modified front-end impedance network, as shown in Fig. 3.6b, can be used to synthesize a single-stage micro inverter with reduced component count [30]. However, such modified topologies have a limited usability when they are to be used in micro inverter application as they have a limited voltage boosting capability.

3.4   GRID-ISOLATED MICRO INVERTERS

The next class of single-phase micro inverters used in distributed PV power conversion systems is load (or grid) isolated inverters with transformers. In order to overcome drawbacks associated with low frequency (LF) transformers, high frequency (HF) transformers are used in isolated micro inverter systems. As categorized in Fig. 3.1, grid isolated micro inverters can have either a DC-link, pseudo DC-link or a high frequency link. The corresponding inverter configurations are shown in Fig. 3.7. At least two power conversion stages are needed with the inclusion of a

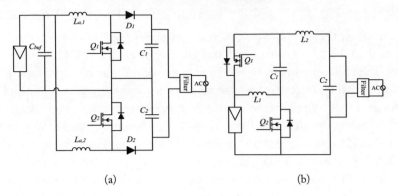

(a)   (b)

Figure 3.6: Transformerless micro inverters with (a) differential mode operation and (b) quasi Z-source.

HF isolation transformer compared to transformerless micro inverters where a single power conversion stage is used. However, due to the isolation of load and source, simple power modulation strategies can be used in these topologies without complex topological modifications to mitigate leakage ground currents. The number of power conversion stages in a particular grid isolated micro inverter depends on the nature and the availability of the DC-link as shown in Fig. 3.7. In

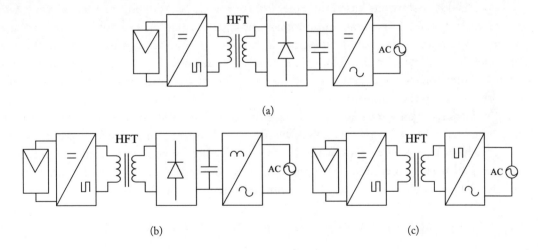

Figure 3.7: Grid isolated micro inverters with (a) DC-link, (b) pseudo DC-link, and (c) high-frequency-link.

the DC-link and pseudo DC-link type micro inverters, there are three power conversion stages whereas DC-link-less type has only two conversion stages. Therefore, the DC-link-less or high-frequency-link (HFL) type micro inverters can be used to improve efficiency and power density of the grid-isolated micro inverters. However, a proper mechanism has to be used to reduce double line frequency power ripple in the front-end DC-bus. The most common approach is to use a sufficiently large capacitor. In order to meet reliability and efficiency requirements, capacitors with high reliability figures and low ESR should be used which in turn increases the cost. Moreover, DC current injection to the utility grid as a result of unbalance operation of the DC-AC inverter needs to be mitigated.

3.4.1 GRID-ISOLATED MICRO INVERTERS WITH A DC-LINK

There are three power conversion stages in DC-link type micro inverters. With reference to Fig. 3.7a, these stages can be identified as the DC/HF AC stage which is at the left side of the HF transformer, the HF AC/DC stage which is at the right of the HF transformer and DC/LF AC stage which is connected to the utility grid. The rectified HF AC output of the second power conversion stage is fed in to the DC-link capacitor of the grid connecting inverter which absorbs the double line frequency power ripple. Therefore, a small film capacitor can be used at the front-end and thereby improve the reliability of the micro inverter system. The voltage boosting capability of the HF isolation transformer reduces the operational complexity of the front-end DC/HF AC converter. Therefore, the main function of the front-end converter can be simplified to the tracking of MPP of the PV module. The grid-side power conversion stage injects a synchronized near sinusoidal current into the utility grid by maintaining the DC-link voltage at a desired level.

The topology and power modulation strategy selection for each power conversion stage in this type of micro inverters needs to be done with extra care in order to minimize the number of semiconductor devices and to reduce power losses. The DC/HF AC converter can be full-bridge, half-bridge, push-pull, or other single-ended type topologies such as fly-back or forward converters as shown in Fig. 3.8. Fly-back and forward converter topologies have reduced number of power switches compared to full-bridge converter. However, such single-ended topologies have a low efficiency due to asymmetrical core utilization of the isolation transformer. However, this drawback may not be a significant problem with low power applications such as single PV module with rated power less than 300W. This issue can be solved using half-bridge and push-pull converter topologies at the expense of higher number of power switches. The rectifying stage of the micro inverter can be designed by taking into consideration the turns ratio of the isolation transformer and the number of semiconductor devices in the rectifier. A single diode can be used with single ended topologies and a full-wave or half-wave rectifier can be used with the other converter topologies. A voltage doubler rectifier can be used together with a half-wave rectifier in order to reduce the turns ratio of the isolation transformer. Full-bridge converters are often used as the grid-connected inverter to generate a grid synchronized current. The switching frequency of the grid-connected inverter depends on the applied power modulation method. In some cases, switching frequency can be reduced in order to reduce switching losses.

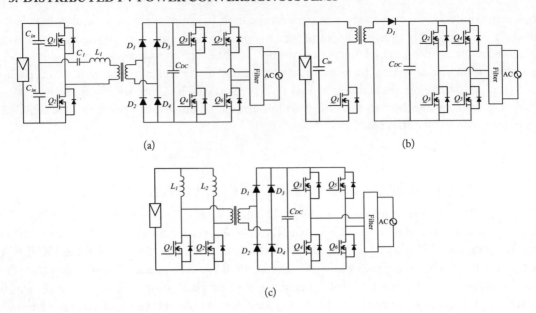

(a)

(b)

(c)

Figure 3.8: Micro inverters with DC-link based on (a) half-bridge converter, (b) fly-back converter, and (c) push-pull converter.

A resonant power converter with a front-end half-bridge converter and grid-connected full-bridge converter is shown in Fig. 3.8a. The full-wave rectifier at the secondary side of the isolation transformer rectifies the HF resonant current flown out of the series resonant circuit at the output of the front-end converter. The front-end half-bridge converter can be replaced by a full-bridge converter in order to reduce the turns ratio of the isolation transformer as voltage conversion ratio of the full-bridge converter is twice that of half-bridge converter. The resonant operation makes half-bridge converter power switches to turn-on with ZVS and turn-off with reduced losses when an external capacitor is connected between the drain and source of the power MOSFET. Moreover, the resonant operation modifies switching waveforms of the full-bridge diode rectifier in a way that the power losses get reduced. One leg of the full-bridge grid-connected converter in Fig. 3.8a is switched at LF in order to reduce switching losses while the switches in the other leg are switched at HF. A micro inverter with a fly-back converter is shown in Fig. 3.8b with a grid-connected full-bridge converter [31]. This micro inverter has relatively fewer number of semiconductor devices compared to the micro inverter shown in Fig. 3.8a but at the expense of higher losses due to high peak and root mean square currents. This drawback can be minimized using interleaved fly-back converters by sharing the input current among the two input power switches. In such a case, the configuration of the isolation transformer can become complex and the number of semiconductor devices increases. A micro inverter with front-end current-fed dual inductor push-pull (or isolated two switch boost) converter is shown in Fig. 3.8c with a grid-connected full-bridge converter [32]. The turns ratio of the HF isolation transformer can be reduced with the active boosting operation of the front-end converter. However, a mechanism to reduce high voltage surges at turn-off of the push-pull converter power switches due to leakage inductance of the isolation transformer needs to be in place. The switching losses can be overcome using the resonant operation between leakage inductance of the high-frequency transformer and output capacitances of the power switches with variable frequency operation.

3.4.2 MICRO INVERTER WITH A PSEUDO DC-LINK

There are three power conversion stages in the isolated micro inverter with pseudo DC-link as shown in Fig. 3.7b. The operation of the front-end converter is similar to the DC-link micro inverter. However, it generates an output with the help of the second power conversion stage that follows the envelope of a rectified 50/60 Hz sinusoid. The output of the second power conversion stage is fed into an unfolding circuit in order to obtain a grid-frequency sinusoidal current. Therefore, the grid-connected converter behaves as a current source inverter (CSI). Two power switches of the CSI can be operated at a low switching frequency and the remaining two switches can be operated at a higher switching frequency in order to reduce switching losses. The passive power decoupling capacitor of this type of micro inverters is placed at the input of the front-end converter and hence the reliability can become an issue if electrolytic capacitors are used. A pseudo DC-link micro inverter with a front-end fly-back converter is shown in Fig. 3.9a [33]. The fly-back converter is controlled to obtain an amplitude modulated output current which follows the envelope of a 50/60 Hz sinusoid. The output current is then fed to the grid-connected CSI to obtain a grid synchronized current. The topology shown in Fig. 3.9a can be modified using a modified transformer configuration and fewer number of semiconductor devices which is known as interleaved fly-back converter. The front-end converter of the pseudo DC-link micro inverter shown in Fig. 3.9a can be replaced with a current-fed push-pull converter as shown in Fig. 3.9b [34]. The rectified output of the push-pull converter is fed into a buck converter to obtain an output current which follows the envelope of the rectified grid voltage. This micro inverter has three power conversion stages and hence there is better controllability in the auxiliary functions of the micro inverter. A front-end full-bridge converter-based micro inverter is shown in Fig. 3.9c [35]. Similar to the operation of the other two topologies, the rectified sinusoidal output of the full-bridge diode rectifier is fed into the grid-connected converter. The front-end full-bridge is controlled to track the MPP and to obtain a resonant current that can be synthesized to form a grid synchronized current. The full-bridge converter generates square or quasi square wave output and it is then fed in to the second order series resonant circuit which acts as filter on its inputs. The phase shift between two legs of the full-bridge converter or switching frequency can be modulated to obtain amplitude modulated resonant current. The resonant operation reduces switching losses of the full-bridge converter and thus improves switching characteristics of

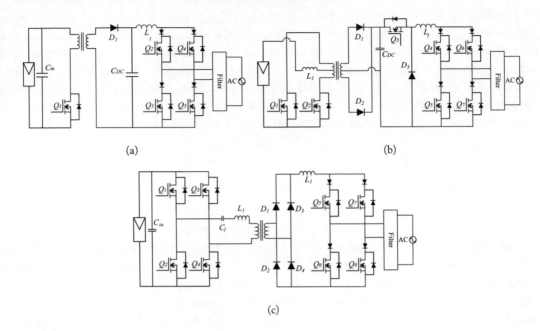

Figure 3.9: Micro inverters with pseudo DC-link based on (a) fly-back, (b) push-pull, and (c) full-bridge converters.

the full-bridge rectifier. The unfolding circuit forms an output current with low distortion at the zero crossing of the grid voltage.

3.4.3 DC-LINK LESS OR HIGH-FREQUENCY-LINK MICRO INVERTERS

There are two power conversion stages in the DC-link-less or high-frequency-link (HFL) micro inverters as opposed to three power conversion stages in the aforementioned DC-link micro inverters. The front-end converter generates a HF voltage in the HFL and it is modulated by the grid-connected converter to obtain a grid synchronized output voltage. The grid-connected converter operates as a frequency changer (also known as cycloconverter) and its operation can be either full-wave or half-wave. The front-end converter of the micro inverter needs to track MPP of the PV module. The complexity of the controller of the micro inverter is reduced to a certain extent due to the absence of the DC-link capacitor. Similar to the micro inverter topologies discussed above a buffer capacitor has to be placed at the input side of the front end converter of these topologies as well in order to mitigate double line frequency power ripple. Therefore, HFL micro inverters suffer from reduced reliability if an alternative power decoupling method is not used. Several alternative power decoupling methods are explained in Chapter 4. A few HFL micro inverters without power decoupling circuits are shown in Fig. 3.10.

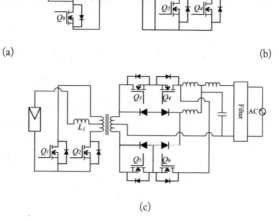

(a)

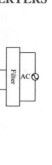

(b)

(c)

Figure 3.10: HFL micro inverters with a (a) half-wave, (b) and (c) full-wave cycloconverters.

Figure 3.10a shows a micro inverter with a front-end full-bridge converter and a grid-connected half-wave cycloconverter. The series resonant circuit (L_1 and C_1) is used to turn-on power switches with ZVS and ZCS and thereby reducing power losses [36]. The ZVS and ZCS operation of the front-end full-bridge converter depends on the ratio between resonant frequency to the switching frequency of the micro inverter. There is ZVS at the turn-on of the full-bridge converter power switches when that ratio is smaller than one and ZCS at the turn-off of the switches when it is greater than one. There is another capacitor (C_2) in the secondary side of the HF isolation transformer to block DC magnetizing current. This DC blocking capacitor needs to be significantly larger than the resonant capacitor in order to prevent interferences between them. Frequency and phase shift modulation-based power control strategies are used with resonant power converter to control the output power. However, frequency modulation techniques exhibit an unpredictable noise spectrum, complex output filter design, and large frequency variation which can be seen as drawbacks compared to phase shift modulation with fixed frequency-based control strategies. Hence, phase shift modulation is commonly used to track MPP of the PV panel and to obtain grid synchronized output current. A micro inverter with a full-wave cycloconverter is shown in Fig. 3.10b [37]. The operation of the micro inverter shown in Fig. 3.10b is very similar to the micro inverter with a half-wave cycloconverter. However, it has extra switches and therefore possess a higher power rating. The cycloconverter power switches are bidirectional and can be realized using back to back connected power MOSFETs. A full-wave cycloconverter implementation is shown in Fig. 3.10c with center tapped winding and reduced active power switches [38]. A current-fed push-pull converter is used as the front-end converter of the micro inverter and hence MPP of the PV panel is tracked by the primary side power switch control signals. The grid connected filter removes HF switching power ripple in the micro inverter output current. The filter requirement is tighter with half-wave operation compared to the full-wave operation of the cycloconverter.

3.5   MICRO INVERTER CONTROL STRATEGIES

Basic control objectives of micro inverters are MPPT, grid synchronization and DC-link voltage control if a DC-link is present. Moreover, the micro inverter controller needs to be designed to achieve a low total harmonic distortion as much as possible with the given hardware and avoid DC current injection to the grid in order to meet the requirements spelt in the respective grid code. Therefore, micro inverter controller design mainly depends on the number of power conversion stages, presence of a DC-link and grid code requirements.

The control structure of micro inverters with a DC-link is shown in Fig. 3.11a. The MPP of the PV module is tracked by the front-end DC-DC converter using the output voltage and current of the PV module. The control variable of the front-end converter is based on the applied power modulation method, namely, PWM, phase shift or frequency modulation. For example, the duty ratio of the DC-DC converter can be calculated using the output provided by the MPP controller as a voltage reference signal to the PWM modulator. However, there are two control loops in input voltage control-based MPPT systems, namely, slow outer current loop and fast inner voltage loop. A phase locked loop (PLL) is used to track the phase angle of the grid voltage and thereby generate a synchronized grid current reference signal which is compared with measured grid current to form control signals for the grid-connected DC-AC inverter. The grid current reference can be seen as the power reference as well under constant grid voltage conditions. Therefore, an increase of the grid current reference means more power transferred to the grid from the DC-link. If the PV module power increases the front end DC-DC converter feeds more power to the DC-link and as a result DC-link voltage increases. The voltage controller sense this change and increases the grid current reference which in turn draws more power from the DC-link to regulate the voltage to the set point. The opposite happens when the PV module power is decreased. The DC-link voltage can be kept at a constant value irrespective of the changes in solar insolation with this controller.

The voltage and current controllers can be implemented using either proportional (P) or proportional-integral (PI) controllers. The P controller-based solutions need feed forward input power to improve the system response. However, such a feed-forward mechanism increases the coupling between the input and the output and may give rise to a highly distorted output current due to input power fluctuations. Such shortcomings can be overcome using PI controller-based voltage and current controllers. However, as the PI controllers operate on DC quantities their

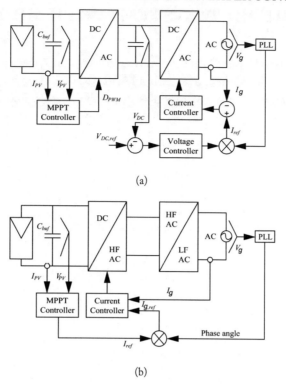

(a)

(b)

Figure 3.11: Controllers of the micro inverter with a (a) DC-link and (b) HFL.

performance becomes poor at high harmonic conditions. As a solution, a different type of controller, known as proportional resonant (PR) controller, has recently been proposed [39]. This controller comprises of a proportional and a resonant control components where the resonant component is tuned to a certain frequency with limited bandwidth. Hence, this controller can be used to mitigate harmonic components in the output current by having several cascaded resonant controllers tuned to different harmonic frequencies.

The controller design of the HFL micro inverters is significantly different from that of the DC-link-based micro inverter discussed above. In HFL micro inverters, both MPP and grid current control are implemented in the front-end converter, as shown in Fig. 3.11b, since the grid-connected converter is used to form a grid-synchronized LF output current. The reference signal of the current controller is synthesized using the PLL output and the output of the MPPT controller. The grid current reference signal is fed into the current controller and is compared with the grid current to generate a control signal for the front-end converter, as shown in Fig. 3.11b.

3.6   DISTRIBUTED PV SYSTEMS WITH MICRO CONVERTERS

Micro converters are used in distributed PV power conversion systems to integrate PV modules to a common DC-bus. The micro converter is a DC-DC converter with high voltage conversion ratio that can be used to connect PV modules in parallel, series, and partial power processing architectures, as shown in Fig. 3.12. The parallel connection of the micro converters to the DC-bus

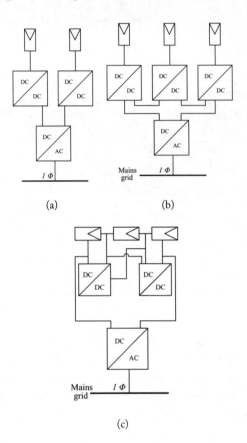

Figure 3.12: Distributed PV systems with micro converters connected in (a) parallel, (b) series, and (c) partial power processing architectures.

provides expansion flexibility, enhanced performance under shading conditions, and less impact on single converter failure. However, micro converters used in parallel connected systems need to possess high voltage conversion ratios compared to series connected systems even though series connected systems do not exhibit the above-mentioned advantages. As a result, isolated DC-DC converter topologies comprising high voltage devices, such as active power switches and diodes, may need to be used in micro converters in parallel connected systems. As power generated by each PV module is processed by an individual micro converter in both parallel and series configurations, the overall power conversion efficiency at the system level can be low.

Partial power processing architecture that processes a fraction of power generated by the PV module is proposed to overcome the aforementioned drawback, as shown in Fig. 3.12c. This gives rise to reduced conduction losses in the DC-DC converters due to their partial power processing. This configuration reduces the volume of the power converters as a single converter processes only a fraction of total power generated by the PV module. Therefore, this is a promising solution that can overcome the above-mentioned drawback of series connected micro inverters. In partial power processing based systems, the DC-DC converter behaves as a current source in string configuration by facilitating the operation of all adjacent PV panels at their MPP while partially processing each PV module generated power. Non-isolated DC-DC converter topologies are commonly used in this configuration as the partial power processing is applicable in series connected PV modules. This idea can be further extended to the sub module level of a PV module to enhance MPPT efficiency using sub module integrated converters (SubMIC).

Hence, both isolated and non-isolated DC-DC converters can be used as micro converters in distributed PV power conversion systems and such micro converters are discussed in the following sections.

3.6.1 DC-DC CONVERTERS WITH ISOLATION TRANSFORMER

High switching frequencies are used in DC-DC converters to increase power density as it is a mandatory requirement in distributed PV systems. As a result, HF isolation transformers have become popular in micro converters owing to their advantages such as compact size, low weight, passive voltage boosting and load isolation. Isolated DC-DC converters can utilize the isolation transformer core asymmetrically to transfer power from primary side to secondary side as in fly-back or forward converters or symmetrically as in full-bridge, half-bridge and push-pull converters. Asymmetrical and symmetrical core utilization can be applied in these converters as they always deal with power generated by a single PV panel and losses due to asymmetrical core utilization may not cause a significant energy loss. However, it is always advantageous to use power converters with reduced number of power switches in order to satisfy basic requirements expected from the micro converters. Hence, micro converter topologies can be based on either push-pull, half-bridge, and single-ended types as opposed to full-bridge topologies. A few such micro converter topologies with HF isolation transformer are shown in Fig. 3.13.

Figure 3.13a shows a micro converter based on resonant half-bridge converter. One of the significant drawbacks of resonant converters is limited operation of the converter in wide input power range due to circulation of energy. This drawback can be overcome by operating the converter in two operation modes based on the output voltage and power of the PV module [40]. However, the number of magnetic components and semiconductor devices is high and it can cause a reduction in the efficiency of the converter. The series connected input to the DC-link

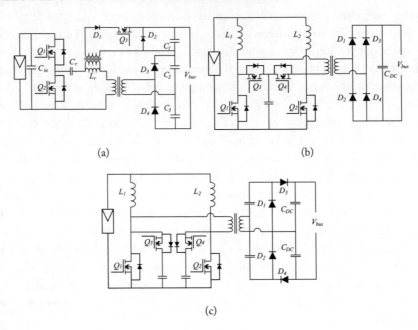

(a)

(b)

(c)

Figure 3.13: Isolated micro converter topologies with (a) half-bridge, (b) current-fed push-pull, and (c) current-fed push-pull converter with voltage quadruple.

increases the voltage conversion ratio. A micro converter topology based on dual inductor current-fed push-pull converter (or isolated boost converter) is shown in Fig. 3.13b with active clamping circuit to reduce voltage stress on active power switches at turn-off due to leakage inductance of the isolation transformer [41]. The stresses on the primary side active power switches can be overcome by using an auxiliary transformer and clamping winding without using active power switches in order to reduce the cost and to increase reliability. The diode rectifier in the secondary side of the isolation transformer can be replaced with an active rectifier in order to obtain ZVS and ZCS at the primary side power switches with modified switching control strategies [42]. Such solutions not only help to improve switching characteristics but also help to obtain higher voltage conversion ratios.

An isolated boost converter with a secondary side voltage quadruple circuit is shown in Fig. 3.13c. This converter exhibits a high voltage conversion ratio and possesses secondary side diodes with reduced voltage stresses. Hence, a low-voltage high-performance diode can be used to improve the efficiency of the converter. Moreover, all primary side power switches can be turned-on with ZVS with the help of auxiliary switches in the primary side.

3.6.2   DC-DC CONVERTERS WITHOUT ISOLATION TRANSFORMER

A micro inverter based on non-isolated DC-DC converter is more suitable in series and partial power processing applications as the required voltage conversion ratios are relatively low compared to parallel connected systems. Hence, simple boost converter topologies and their derivatives with improved features such as high voltage conversion ratio, reduced input current ripple and high efficiency can be used as micro converters, as shown in Fig. 3.14. The high voltage gain of such non-isolated topologies are obtained using techniques such as cascaded structures, switched inductors, switched capacitors, voltage lift, coupled inductors, and front-end impedance networks.

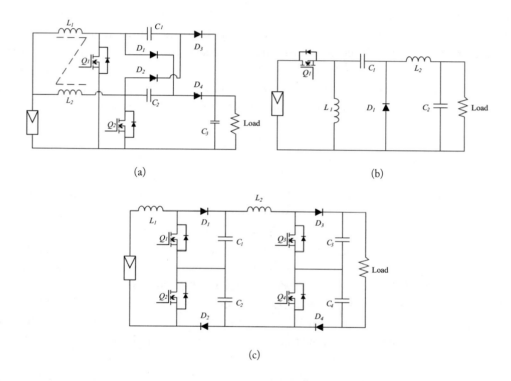

Figure 3.14: Non-isolated micro converters with (a) coupled inductor, (b) voltage lift, and (c) cascaded three-level topologies.

A micro converter with coupled inductors and switched capacitors is shown in Fig. 3.14a. This configuration helps to improve magnetic component utilization with the use of a coupled inductor as opposed to separate inductor-based interleaved configuration. Moreover, a high voltage conversion ratio is obtained using a switched capacitor (C_1 and C_2) network. A DC-DC converter based on voltage lift technique to obtain high voltage conversion ratio is shown in Fig. 3.14b. The capacitor (C_1) helps to lift the input voltage to a desired level by transferring the energy in the inductor (L_1) stored during the turn-on time of the active power switch. The voltage gain obtained using the voltage lift structure can be further increased using several cascaded voltage lift cells. A non-isolated step-up converter based on three-level boost converters is shown in Fig. 3.14c. The boost converter with multi-level output possesses high voltage conversion ratio with reduced voltage stresses on active devices. However, power switches are turned on and off without soft switching and diodes have high reverse recovery losses.

3.7 SUB-MODULE INTEGRATED CONVERTERS

The distributed PV power conversion systems discussed so far are based on converter per module architecture. However, the most common issues such as shading and module mismatch can be present even at the module level. The sub-module integrated converters are proposed to overcome those drawbacks using separate bidirectional DC-DC converters connected to each sub-string comprising of series connected PV cells in a PV module. As a result, MPP of each PV sub-string in a PV module can be tracked more precisely at a basic level. The sub module integrated converters can be connected in a similar manner to that of per module converters, in series (sigma) and partial power processing (delta) architectures. In sigma configuration, PV cell string is connected in parallel to the DC-DC converter and the output of all converters are connected in series in the sigma configuration. Hence, DC-DC converters need to possess highest efficiency to obtain the maximum power at the output of the PV panel. However, in the delta configuration, all PV cell strings are connected in series and DC-DC converters are connected in partial power processing architecture. Therefore, MPPT of PV modules becomes complex in delta configuration due to the presence of several MPPs in the characteristic curve compared to the sigma converter where the converter can only see its own cell characteristics and the corresponding MPP.

Non-isolated DC-DC converter topologies are typically used as SubMIC in order to obtain high efficiency and power density. The required voltage step up capabilities are obtained using similar techniques, as in non-isolated DC-DC converters explained in Section 3.6.2, but with a significantly high switching frequency to reduce the volume of passive devices.

3.8 POWER CONVERTER TOPOLOGY ANALYSIS

A few power converter topologies that can be used in micro inverter and micro converter applications are discussed in Sections 3.3–3.6. The front-end converter of these topologies can be voltage-fed or current-fed type depending on the filter inductor placement and power modulation methods. Current-fed topologies have a significant advantage over voltage-fed topologies as a result of their low-input current ripple and high-voltage conversion ratio owing to the boost nature of the topology in low-input voltage and high-input current applications. Moreover, there is an advantage in MPPT algorithm developed using input voltage control MPP methods as they have better performance over input current control-based methods.

HF isolation transformers are used in both micro inverter and converter applications to obtain load isolation and passive voltage boosting. Such HF transformers can be single-input–single-output with two windings or dual-input-dual-output with four windings. Furthermore, transformers based on three windings or center-tap winding with either two terminals in the primary side or secondary side are also used in some converter configurations.

Resonant circuits are widely used in power converters aiming to reduce switching losses in power semiconductor devices. The resonant components can be realized using leakage inductances of the circuits, output capacitances of the devices in addition to using external resonant inductors, and capacitors. A variable switching frequency is used in such topologies to modulate the output power and obtain soft-switching operation in a wide operating range. However, this is not feasible in some applications due to difficulties in designing EMI and output filters for converters due to the unpredictable noise spectrum of the variable frequency operation.

3.8.1   SOFT-SWITCHING IMPLEMENTATION OF POWER CONVERTER

There are two types of soft-switching methods that are commonly used in HF power converters namely zero-voltage-switching (ZVS) and zero-current-switching (ZCS). The implementation of these soft-switching methods completely depend on the power converter topology and power modulation method. The power modulation method needs to facilitate the basic requirement for ZVS and ZCS operations as illustrated in Fig. 3.15.

In hard switching, active power semiconductor switch current and voltage waveforms are overlapped at turn-on and turn-off instances as shown in Fig. 3.15a due to several factors such as output capacitances of power switches. Therefore, hard-switching gives rise to significant power loss during switching which is known as switching loss. Since the number of overlapping is proportional to the switching frequency, switching loss is high at high frequencies and it can even dominate the conduction loss if the switching frequency is beyond certain limit. This drawback can be overcome by forcing the voltage across the power switch to a value near zero before turn-on of the power switch or forcing the current through power switch to a value near zero by diverting it through an anti-parallel body diode as shown in Fig. 3.15b. Such behavior of power converters is not natural and it is influenced by modified switching control strategies and power converter topologies as explained below. However, the cost of topology modifications should be reasonable in order to maintain indices of the performance matrix of power converters to an acceptable level.

Power switches of the full-wave cycloconverter-based micro inverter shown in Fig. 3.10b experience ZVS during turn-on due to series resonant circuit and phase shift power modulation. This can be further explained using waveforms and equivalent circuits shown in Fig. 3.16. The negative resonant current during time interval t_0 to t_1 flows through power switches, as shown in Fig. 3.16a. The same set of power switches conduct the resonant current with the reversal of its direction during time interval t_1 to t_2, as shown in Fig. 3.16b. The power switches U_7 and U_9 are turned-off at the end of this time period. However, the secondary side current should be continuous due to the current source nature of the series resonant circuit. Hence, current flow is

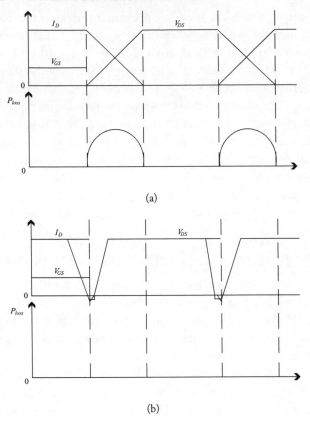

Figure 3.15: The switching waveforms that needs to explain (a) hard and (b) soft-switching operations.

diverted through anti-parallel diodes of U_5 and U_{11}, as shown in Fig. 3.16c, as power switches U_6 and U_{12} are already turned-on during positive half cycle of the grid voltage. This causes discharging of output capacitances of the power switches U_5 and U_{11} during time period t_2 to t_3. As a result, U_5 and U_{11} can be turned-on with zero voltage at the beginning of time period t_3 to t_4. The ZVS turn-on of the power switches has a significant impact from dead time between power switch turn-off and turn-on in order to guarantee ZVS turn-on at all switching conditions. The ZVS operation of the power switches can be further analyzed using drain to source current (I_{DS}) and voltage waveforms, as shown in Figs. 3.17 and 3.18.

The drain to source current of U_5 becomes positive before the gate control signal is applied, as shown in Fig. 3.17. The positive current flows through anti-parallel diode of the power switch. This anti-parallel diode can be either the body diode of power MOSFET or an external diode with low reverse recovery time. As a result, voltage across the power switch reduces to forward

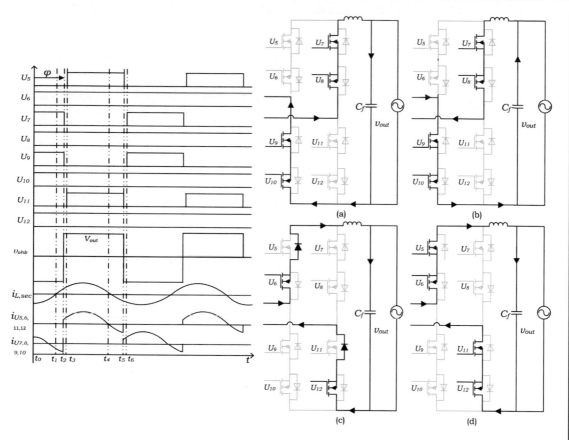

Figure 3.16: Switching control signals and equivalent circuits of the full-wave cycloconverter when power switches U_5 and U_{11} are to be turned-on with ZVS.

voltage drop of the diode before the gate control signal is applied. Hence, the power switch is turned on with ZVS as shown in Fig. 3.18.

Such power modulation methods turn-on full-wave cycloconverter power switches with ZVS at all grid voltage values as the basic idea behind realization of ZVS is based on the diversion of resonant current through anti-parallel diode before turn-on of the power switches. This behavior of back-end converter of the micro inverter is supported by the three-stepped quasi-square wave output voltage generated by the front-end full-bridge converter using phase shift modulation. The phase shift modulation and resonant operation turn-on the full-bridge converter power switches at zero voltage.

The above example illustrates the use of feasible power processing topologies and modulation strategies to obtain soft-switching at different stages of power converters. This strategy can be

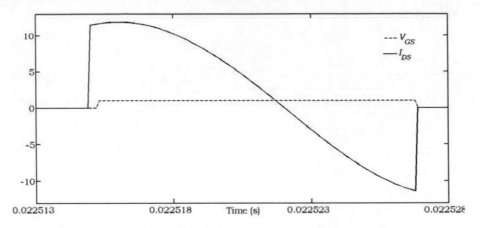

Figure 3.17: Drain to source current (I_{DS}) of power switch U_5 with its gate control signal (V_{GS}).

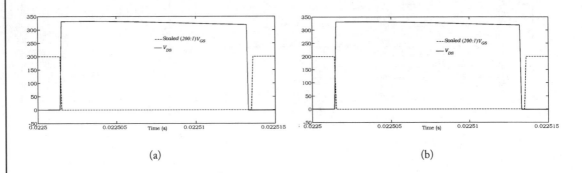

Figure 3.18: Drain to source voltage (V_{DS}) and scaled (200:1) gate control signal of power switches (a) U_5 and (b) U_{11}.

further explained using a ZVS and ZCS current-fed push-pull (CFPP) converter that can be used as micro converter in distributed PV power conversion systems. There are two forms of CFPP converters that can be identified based on the number of inductors at the input of the converter, as shown in Fig. 3.19. They are named as dual inductor and single inductor CFPP converters and they have their own advantages and disadvantages as explained in [43]. The diodes of the voltage doubler rectifier or rectifier with center tap winding at the secondary side of the isolation transformer can be replaced with active power switches, as shown in Fig. 3.19. A switching control strategy can be proposed for these converters in order to turn on and turn off the primary side power switches with ZVS and ZCS, respectively. ZVS and ZCS of the primary side power switches can be explained using switch control signals and ideal current waveforms shown in Fig. 3.20 for single inductor CFPP converter.

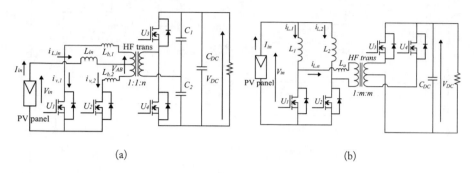

(a)   (b)

Figure 3.19: CFPP converter with a, (a) dual inductor front-end and voltage double rectifier at the back-end and (b) single inductor primary and center-tapped secondary winding.

The primary side power switches of the single inductor CFPP converter is switched using signals with duty ratio larger than 0.5. Such a control signal is applied during t_0 to t_1 as shown in Fig. 3.20 to switch U_2. The anti-parallel diode of U_3 transfers power to the load during this period as shown in Fig. 3.20a. The secondary side power switch U_4 is turned-on during t_1 to t_2 as shown in Fig. 3.20b in order to forward bias the anti-parallel diode of U_1 as it is to be turned-on at the beginning of time interval t_2 to t_3. Subsequently, power switch U_1 is turned-on with ZVS at $t = t_2$. The negative current flows for a short period of time as shown in Fig. 3.20c due to the line inductance although the secondary side power switch U_4 is turned-off at $t = t_2$. Consequently, the current conduction through power switch U_1 changes its direction as shown in Fig. 3.20d.

The ZCS turn-off operation of the single inductor CFPP converter power switches can be explained using Fig. 3.21 and ideal current waveforms are shown in Fig. 3.20 with switching control signals.

The secondary side power switch U_3 is turned on before turn off of the primary side power switch U_2, as shown in Fig. 3.21a. Turn-on of the secondary side power switch makes the anti-parallel diode of the power switch to be forward-biased due to scaled (based on the turns ratio of the isolation transformer) secondary side voltage, as shown in Fig. 3.21b. Then power switch U_2 can be turned-off with ZCS at the end of time interval t_5 to t_6. The negative current continues for a short time period, as shown in Fig. 3.21c with the presence of line inductances $L_{b,1}$ and $L_{b,2}$. The ZVS and ZCS operations are common to all primary side power switches and can be further analyzed using waveforms shown in Fig. 3.22.

Figure 3.22a shows drain to source voltage (V_{DS}) and gate control signal of power switch U_1 when it is to be turned on and turned off. V_{DS} reaches zero before the gate control signal is applied and it remains at that level after turn off of power switch. The prior case directly validates ZVS turn-on of the power switch and later case confirms the current conduction through the

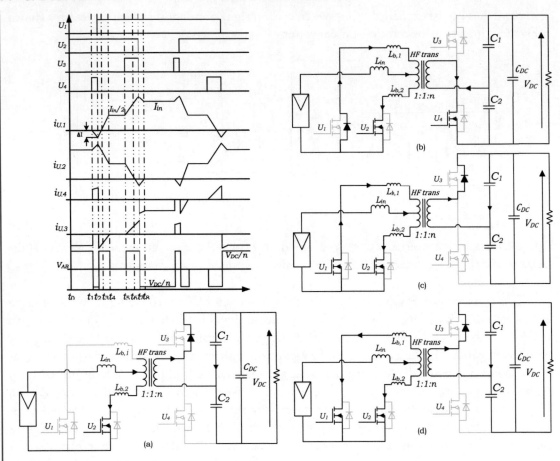

Figure 3.20: Switching control signals, ideal waveforms and equivalent circuits of the single inductor CFPP when power switch U_1 is to be turned-on.

anti-parallel diode of the power switch after its turn-off. ZCS turn-off operation of the power switch can be further validated by analyzing power switch current, as shown in Fig. 3.22b. The power switch current is negative before the gate control signal is applied with the conduction of forward biased anti-parallel diode as explained above.

The ZVS and ZCS operations of the primary side power switches of the converter are facilitated by the feedback of secondary side power to the primary side in order to alternate current conduction through power switches before the primary side switches are turned-on and turned-off. The turn-on time period of the secondary side power switches should be minimized in order to improve efficiency. However, this has a negative impact on the additional voltage gain obtained as the voltage gain becomes a function of the secondary side power switch turn-on time with the

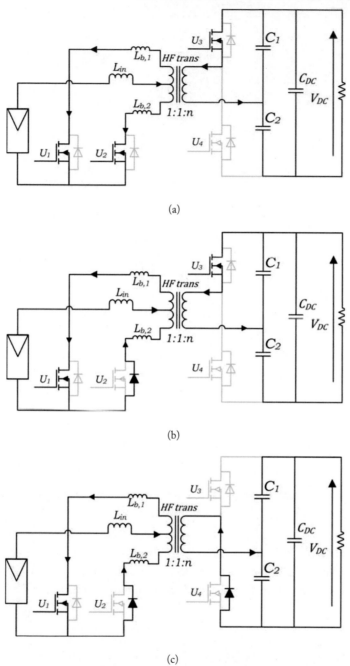

(a)

(b)

(c)

Figure 3.21: The equivalent circuits of the single inductor CFPP converter when power switch U_2 is to be turned off.

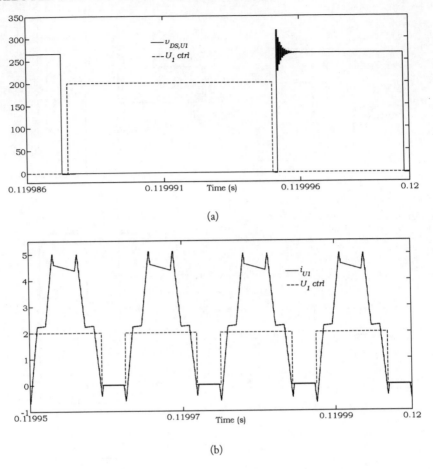

Figure 3.22: (a) The drain to source voltage (V_{DS}) and (b) drain current of power switch U_1 with gate control signal.

aforementioned switching control strategy. Moreover, similar switching control strategy can be applied for dual inductor CFPP converter without modifications to obtain ZVS turn-on and ZCS and turn-off irrespective of the secondary side rectifiers shown in Fig. 3.19.

CHAPTER 4

Active Power Decoupling in Single-Phase Micro Inverters

4.1 INTRODUCTION

Photovoltaic cells can be electrically modeled as of having a DC current source front end. This power source ideally delivers a flow of DC power to the load. Therefore, there is no reactive power associated with PV cells. Moreover, reverse power flow is not accepted and not allowed in most systems. However, the single- phase or three-phase power grids, which in most cases act as the load for PV systems, are alternating current (AC) systems. Particularly, in single-phase AC power systems, the active power is a combination of a direct current (DC) quantity and an AC quantity. The frequency of the AC quantity is twice the fundamental AC frequency. This is known as the double-line frequency, as mentioned earlier. When an inverter is used to interface a PV system, which is a DC source, to a single-phase AC grid, the inverter output power waveform is a DC quantity superimposed with the double-line frequency. In other words, current drawn from the PV cell is no longer steady instead it is a DC quantity superimposed with the double-line frequency. The maximum power extraction from PV cells becomes tedious with this double-line frequency current ripple.

A large electrolytic capacitor placed in parallel with the PV cell can help reduce this current ripple and thereby perform the maximum power point tracking without any difficulty. However, in this case current ripples are diverted to the DC-bus capacitor. The lifetime of electrolytic capacitors is significantly reduced when they are exposed to this type of current ripples. As a result, capacitor failures become the most dominant mode of failures in PV systems. These current ripples can even lead to catastrophic failures in electrolytic capacitors. As a consequence, power converters fail to match with the usual 25-year lifetime guarantee of PV cells. Moreover, increase in the DC-bus capacitance adds significant cost to the power converter.

The double-line-frequency power ripple can be observed in the front-end DC-bus of the single-phase HFL and pseudo DC-link-based micro inverters as well. Therefore, suitable measures have to be taken to reduce it irrespective of the power converter topology used. One alternative is to replace electrolytic capacitors with other types of capacitors such as film capacitors that can withstand high temperatures, voltage and current ripples. However, their power density and cost are not in an acceptable level for them to be used in micro inverters. Hence, there should be a proper mechanism to reduce such voltage and current ripples in micro inverters to improve

MPPT capability without sacrificing the reliability and significantly influencing their cost, power density, and efficiency.

4.2   SINGLE-PHASE OPERATION OF MICRO INVERTERS

The basic configuration of micro inverters that can be used to understand double-line-frequency power ripple generation in the front-end DC-bus with pseudo DC and HF links are shown in Fig. 4.1. In pseudo DC-link type micro inverters the front-end converter generates a rectified voltage at grid frequency and it is unfolded by the grid-connected unfolding circuit to obtain grid synchronized output voltage. In HF link-type micro inverters, the front-end converter generates a HF AC voltage output and it is converted by the grid connecting cycloconverter to obtain a grid synchronized output. Both micro inverter topologies decouple the double-line frequency power ripple using buffer capacitor (C_{buf}) at the front-end DC-bus, as shown in Fig. 4.1.

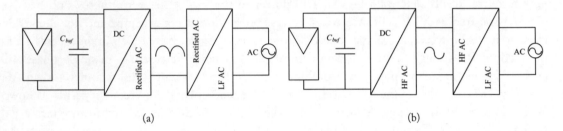

(a)   (b)

Figure 4.1: The micro inverters with (a) pseudo DC and (b) HF links.

The size of the buffer capacitor can be estimated considering the grid voltage and current in the form given in (4.1).

$$v_g(t) = V \sin(\omega_g t)$$
$$i_g(t) = I \sin(\omega_g t + \varphi),$$

(4.1)

where V and I are peak values of the gird voltage and current, respectively, and ω_g is angular frequency of the grid voltage. The phase shift between grid voltage and current is given by phase angle (φ) and it should be kept at zero in practical applications to have unity power factor. The instantaneous output power of the micro inverter is given by (4.2) with phase shift. The instantaneous output power of the micro inverter with unity power factor is given by (4.3).

$$P_g(t) = \frac{VI}{2}\cos(\varphi) - \frac{VI}{2}\cos(2\omega_g t + \varphi)$$

(4.2)

$$P_g(t) = \frac{VI}{2} - \frac{VI}{2}\cos(2\omega_g t), \quad P_g(t) = P_{PV}(t) + P_{AC}(t).$$

(4.3)

The power decoupling operation of a micro inverter can be further analyzed using the simplified expression given by (4.3). The instantaneous output power of the micro inverter consists of two components as given by (4.3). The constant power component (P_{PV}) is provided by the PV module. The pulsating power component at the double-line frequency shown in Fig. 4.2 needs to be handled by the front-end DC-bus capacitor.

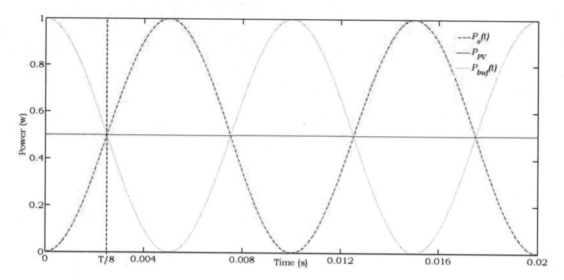

Figure 4.2: Power generated and processed by different elements of the micro inverter.

The pulsating output power ($P_o(t)$) of the micro inverter is depicted in Fig. 4.2 with PV-generated constant DC power (P_{PV}). The power difference between these two elements needs to be supplied by the buffer capacitor. As a result, buffer capacitor voltage is varied at double line frequency within its maximum (V_{max}) and minimum (V_{min}) values. The relationship between energy stored in the buffer capacitor and its maximum (V_{max}) and minimum (V_{min}) voltages is given by (4.4).

$$E_{buf} = \frac{1}{2} C_{buf} \left(V_{max}^2 - V_{min}^2 \right). \tag{4.4}$$

The energy stored in the front-end DC-bus buffer capacitor is equal to difference between PV generated energy and energy injected to the utility grid as given by (4.5) within time period 0– $T/8$, as shown in Fig. 4.2:

$$E_{buf} = \int_0^{\frac{T}{8}} \left(P_{PV} - P_g(t) \right) dt = \frac{P_{PV}}{2\omega_g}. \tag{4.5}$$

The capacitance of buffer capacitor can be estimated using relationships given by (4.4) and (4.5) as given in (4.6). The expression for the buffer capacitor is obtained assuming PV-generated power is equal to the average power term (P_{PV}) in (4.3). The front-end DC-bus buffer capacitance is a function of the output power of the PV module (P_{PV}), DC-bus voltage (V_{DC}), and allowable maximum voltage ripple (ΔV_{DC}). The relationship can be further analyzed using Fig. 4.3 which depicts variation of the buffer capacitance with different parameters.

$$C_{buf} = \frac{P_{PV}}{2\pi f_g V_{DC} \Delta V_{DC}}. \tag{4.6}$$

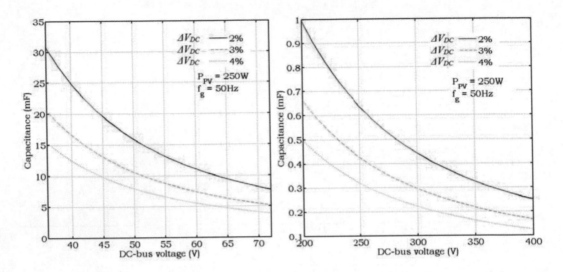

Figure 4.3: The front-end DC-bus capacitances vs. voltage with different DC-bus voltages and voltages ripple percentages.

The required DC-bus capacitances for different DC-bus voltages is compared in Fig. 4.3 calculated using (4.6) at a constant PV generated power (250W) and grid voltage frequency (50 Hz) with different DC-bus voltage ripple percentages. The capacitance requirement for lower allowable voltage ripple in the DC-bus is high compared to that of higher voltage ripple. A similar behavior can be observed in a different voltage range, as shown in Fig. 4.3. Also, there is a significantly reduced capacitance requirement at high DC-bus voltages and therefore, it can be viewed as a feasible option to mitigate double line frequency power ripple by placing a passive power decoupling capacitor at the high voltage DC-link if it is available as explained in Section 4.4.

The DC-bus voltage (V_{DC}) and its voltage ripple are calculated by averaging minimum and maximum voltages of the front-end DC-bus as given by (4.7):

$$V_{DC} = \frac{V_{max} + V_{min}}{2}, \quad \Delta V_{DC} = V_{max} - V_{min}. \tag{4.7}$$

An expression for maximum allowable voltage ripple can be obtained assuming double-line frequency voltage ripple in the DC-bus as given by (4.8):

$$v_{DC}(t) = V_{DC} + V_r \sin(2\omega_g t), \quad V_r = \frac{\Delta V_{DC}}{2}. \tag{4.8}$$

Then current through the buffer capacitor is given by (4.9):

$$i_{buf} = C_{buf}\frac{dv_{DC}(t)}{dt} = 2\omega C_{buf}V_r \cos\left(2\omega_g t\right). \qquad (4.9)$$

The power processed by the buffer capacitor can be calculated using (4.10):

$$p_{buf} = 2\omega C_{buf}V_r\left(V_{DC} + V_r \sin\left(2\omega_g t\right)\right)\cos\left(2\omega_g t\right). \qquad (4.10)$$

The energy stored in the buffer capacitor is given by (4.11) and (4.12):

$$E_{buf} = \int_0^{\frac{T}{8}} 2\omega C_{buf}V_r\left(V_{DC} + V_r \sin\left(2\omega_g t\right)\right)\cos\left(2\omega_g t\right)dt \qquad (4.11)$$

$$E_{buf} = \frac{C_{buf}V_r}{2}\left[2V_{DC} + V_r\right]. \qquad (4.12)$$

An expression of the voltage ripple (V_r) in the DC-bus can be obtained by equating two expressions of stored energy in the buffer capacitor given by (4.5) and (4.12):

$$V_r = -V_{DC} + \sqrt{V_{DC}^2 + \frac{2P_{DC}}{\omega C_{buf}}}. \qquad (4.13)$$

The voltage ripple on the DC-bus is a function of buffer capacitance (C_{buf}) and PV generated power. There is higher ripple in the DC-bus for lower capacitor values as there is inverse relationship between DC-bus voltage ripple and buffer capacitance. This can be further explained using Fig. 4.4 which is drawn using (4.13).

Figure 4.4 shows that voltage ripple in the DC-bus reduces with increased DC-bus capacitance at a particular DC-bus voltage. In other words, the required capacitance to obtain a particular voltage ripple at the DC-bus reduces with increased DC-bus voltage. Moreover, the graphs in Fig. 4.4 indicate that there should be a large capacitor in the DC-bus for low DC-bus voltages (such as 36 V or 72 V). A large electrolytic capacitor bank is required to fulfill this capacitance requirement which is not feasible with micro inverters that need be made as compact as possible.

4.3   POWER DECOUPLING METHODS

The MPPT of micro inverters are based on front-end DC-bus voltage and input current measurements of the micro inverter. Therefore, a proper mechanism is required to mitigate double-line-frequency ripple in the front-end converter voltage and current and thereby perform MPPT accurately. Hence, power decoupling capacitor of pseudo and HF links micro inverter can only be placed at front-end DC-bus. However, such capacitance need to be very high as front-end DC-bus voltage usually ranges from 36–72 V, as shown in Section 4.2. As a result, electrolytic capacitors are needed to fulfill this capacitance requirement in order to keep volume and cost

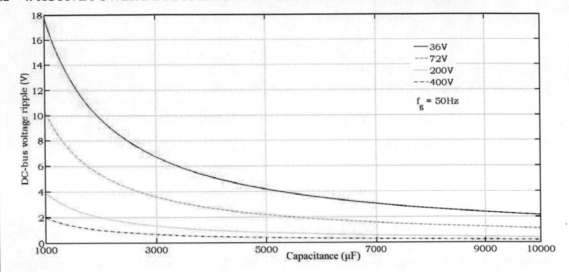

Figure 4.4: DC-bus voltage ripple vs. capacitances at different DC-bus voltages at 50 Hz grid voltage and 250W PV module output power.

of the capacitor bank at economically feasible level. This drawback can be overcome using two options which are based on high voltage DC-links. This classification is based on the strategies utilized to obtain a high voltage DC-link in micro inverters and they are known as active power decoupling and power decoupling using high voltage DC-link capacitor.

4.4 POWER DECOUPLING USING HIGH VOLTAGE DC-LINK CAPACITOR

This is a passive power decoupling method and it does not require an additional power conversion stage to integrate power decoupling capacitor to the micro inverter. As a result, there will be no extra power losses and complex control strategy in the micro inverter. The passive power decoupling capacitor is placed in a medium voltage DC-link of micro inverters with two or more power conversion stages. The front-end DC-DC converter generates a medium voltage DC output within the range of 200–400 V using a PV module output voltage of 36–72 V. Hence, the DC-DC converter need to possess a high voltage conversion ratio. Furthermore, the DC-DC converter can be either isolated or non-isolated type. The DC-link capacitance needed to mitigate double-line-frequency power ripple can be calculated as explained in Section 4.2. The required capacitance is highly dependent on the DC-link voltage and the allowable maximum voltage ripple. The instantaneous DC-link voltage cannot be allowed to go too low as large voltage oscillation may result in malfunctioning of the grid-connected DC-AC converter and distortions in micro inverter output current. However, the connection of a large DC-link capacitance causes an extra cost and volume to the micro inverter. Hence, some research work has been carried out to modify the available DC-link voltage control methods to further reduce the DC-link capacitor size by allowing it to handle higher voltage ripple without degrading the quality of the micro inverter output current.

A DC-link voltage ripple estimation-based voltage controller is proposed for a grid-connected single phase micro inverter in [44] to overcome slow dynamic performance of the bandwidth limited solutions and this converter allows 20% voltage ripple in the DC-link. This solution helps to mitigate negative effects of bandwidth limited DC-link voltage controller, for example in the case of sudden variation in the active power generated by the system. Moreover, notch filters can be used to filter out the second harmonic and other dominant harmonics [45]. This solution improves the dynamic performance and gives rise to high efficiency and power density due to the significant reduction in the DC-link capacitance.

4.5 ACTIVE POWER DECOUPLING (APD)

Active power decoupling methods are used to reduce the DC-link capacitance in the process of mitigating double-line frequency power ripple in single-phase micro inverters. A possible APD scheme is to add an extra power port in the micro inverter with HFL to integrate a small capacitor to the main power circuit. However, this additional circuit reduces power density and efficiency of the power converter. With the addition of the third port the overall efficiency of the micro inverter η_o drops to $\eta_o - 2/\pi(1 - \eta_t)$ where η_t is the efficiency of the additional port. The fundamental idea of using the third power port for power decoupling can be explained using Fig. 4.5. The micro inverter provides power to the utility grid while storing energy in the capacitor of the ripple port as shown in Fig. 4.5a when PV module generated power is greater than the output power. The injected energy to the grid and stored energy in capacitor are represented by areas "A" and "B," respectively, as shown in Fig. 4.5c. Energy stored in the capacitor is released to the grid in parallel with the energy generated by PV module when PV-generated power is lower than the load demand. The PV-generated energy and energy provided by the ripple port capacitors are represented by areas "C" and "D," respectively, as shown in Fig. 4.5c. If the power losses in the ripple port is neglected, the net energy stored in the ripple port capacitor becomes zero when the whole system satisfies the condition, $P_g = P_{DC} + P_{ripple}$ where P_g is the power injected into the grid, P_{DC} is the power extracted from the PV module from the grid and P_{ripple} is the ripple power which is handled by the third port.

The ripple port concept shown in Fig. 4.5 has been realized using different strategies listed below and their basic configurations are shown in Fig. 4.6.

- A power converter connected in parallel with PV module in the DC-bus.

- A power converter connected in series with the power flow.

- A third port connected to the isolation transformer.

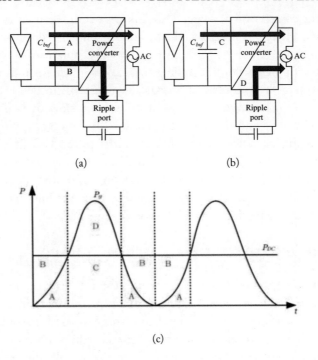

(a)   (b)

(c)

Figure 4.5: Power flow of the micro inverter with an APD circuit.

- A power converter in the AC side of the micro inverter.

Any of these solutions can be used to realize the third power port. However, it adds extra active and passive devices to the micro inverter causing reduced efficiency and reliability and extra cost. Each realization method has its own advantages and disadvantages as explained in the following subsections.

4.5.1   A PARALLEL POWER PORT WITH PV MODULE

The basic configuration of a micro inverter with third power port in parallel with PV module is shown in Fig. 4.6a. The third power port can process either the whole power generated or portion of power generated by the PV module. In either case the third power port is a bi-directional power converter which can be operated as a buck and boost converter. A few such micro inverters with a third power port are shown in Fig. 4.7.

The micro inverter shown in Fig. 4.7a has a bidirectional power port in parallel with the PV module. The APD circuit operates as a boost converter in order to store surplus power generated by PV module and buck converter when stored power is released in to the main power circuit. The overall controller of the micro inverter needs to track the reference current waveform in

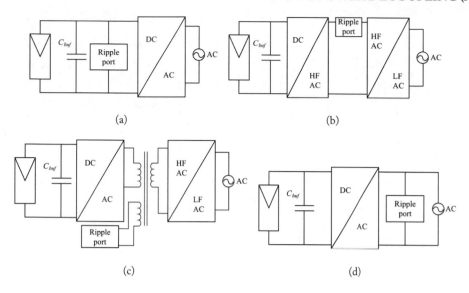

Figure 4.6: Basic configurations of the micro inverters with APD circuits.

order to store and release energy in the power decoupling capacitor. Thus, the power decoupling capacitor partially processes energy generated by the PV module. However, the overall efficiency of the micro inverter is still low due to the losses in the power switches as parallel capacitors are connected across them. A micro inverter comprising a fly-back converter an APD circuit is shown in Fig. 4.7b. The PV-generated power is stored in the power decoupling capacitor, C, using the fly-back transformer of the converter and then the stored energy in the capacitor is released to the grid in synchronization with the sinusoidal grid voltage. Hence, the total power generated by the PV module is processed by two power conversion stages before it is fed it into the grid and as a result the power conversion efficiency becomes lower. The topology shown in Fig. 4.7b is modified as shown in Fig. 4.7c in order to process surplus power generated by the PV module. The PV-generated power is modulated using two power modulation methods known as sequential magnetizing modulation (SMM) and time-shared magnetizing modulation (TMM) [46]. In all three configurations shown in Fig. 4.7 the ripple port, and thus the power decoupling capacitor, is connected to the primary side of the isolation transformer which operates at a lower voltage compared to the secondary side. Therefore, as shown in Fig. 4.3 a large capacitor is required to achieve a low ripple percentage.

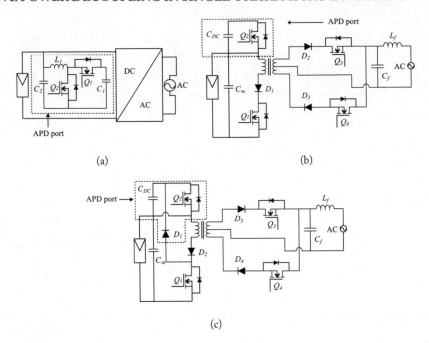

Figure 4.7: Micro inverters with APD circuit in parallel to PV module.

4.5.2   A POWER CONVERTER CONNECTED IN SERIES WITH THE POWER FLOW

The basic configuration of a micro inverter with a series power buffer is shown in Fig. 4.6b. The series power buffer can be placed in HFL of the micro inverter which consists of two power conversion stages and a HF isolation transformer. The series power buffer can be placed either in the primary or secondary side of the isolation transformer. However, the placement in the secondary side of isolation transformer may be more advantageous due to secondary side high voltage and the possibility of varying terminal voltage of the decoupling capacitor in a wider range. A HFL micro inverter with a series power buffer is shown in Fig. 4.8. The operation of the micro inverter is based on the phase shift and frequency modulation. The micro inverter power output is controlled using both methods and the operation of the series power buffer is based on phase shift modulation. The series power buffer modulates the surplus power generated by the PV module and there are increased conduction losses due to the additional components. This micro inverter topology is reported to have a peak efficiency around 98% [47].

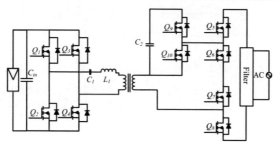

Figure 4.8: Micro inverter with series power buffer.

4.5.3   A THIRD PORT CONNECTED TO THE ISOLATION TRANSFORMER

The basic configuration of a micro inverter with a third power port to mitigate double-line frequency power ripple is shown in Fig. 4.6c. The third bidirectional power port is connected to the micro inverter through a HF isolation transformer having an auxiliary winding. This micro inverter main power flows through power conversion stages of DC-HFAC, HFAC-LFAC or DC-HFAC, HFAC-DC, DC-LFAC. Such a configuration can be easily adopted in a HF isolation transformer-based micro inverter to significantly reduce required capacitance due to possible high voltage conversion ratio that can be obtained using the transformer. This configuration also allows the voltage across power decoupling capacitor to vary without having adverse effect on MPPT or the quality of the output grid current.

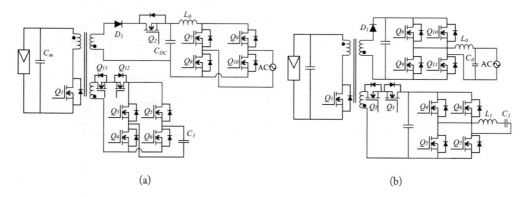

(a)   (b)

Figure 4.9: Micro inverter with auxiliary winding in isolation transformer.

Two such micro inverter topologies with auxiliary winding are shown in Fig. 4.9. Both micro inverters are based on power converters with minimal number of power switches in the main power converter due to additional semiconductor and passive devices of the third power port. The power decoupling capacitor connected to the micro inverter processes surplus energy generated by the PV module. But the efficiency of this converter is low due to the increased number of semiconductor devices and imperfect coupling between magnetics used in integrating the power decoupling capacitor.

The efficiency reduction with the use of auxiliary winding in the transformer can be overcome using an integrated design with only two terminals in the isolation transformer. The third power port can be directly integrated into the secondary side terminals of the isolation transformer.

4.5.4 A POWER CONVERTER IN THE AC-SIDE OF THE MICRO INVERTER

The basic configuration of a micro inverter with an AC side power decoupling circuit is shown in Fig. 4.6d. The size of the power decoupling capacitor can be reduced significantly due to the high voltage and higher allowable voltage ripple across the power decoupling capacitor. Furthermore, this configuration reduces the complexity of the HF isolation transformer and hence the converter efficiency can be improved with its lower dependency on the effectiveness of magnetic coupling. A micro inverter based on current-fed push-pull converter with AC side power decoupling circuit is shown in Fig. 4.10 [48].

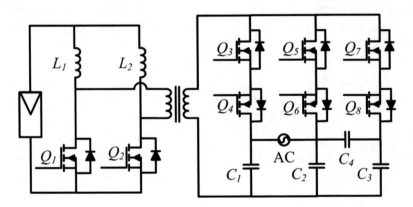

Figure 4.10: Micro inverter with AC side active power decoupling.

The power decoupling capacitor (C_4) is integrated in to the output stage of the micro inverter using bi-directional power switches. This configuration reduces the number of power semiconductor devices and passive devices. Therefore, reliability can be improved as a result of simple driving and control circuits.

4.6   ACTIVE POWER DECOUPLING USING INDUCTIVE ELEMENTS

Power decoupling methods discussed so far use capacitive elements as energy storing devices. One of the major drawbacks of capacitive elements is their limited life time at elevated temperatures as explained in Section 4.1. Inductive elements can also be used as energy storing elements with their advantage of only 0.5% reduction in inductance across over 20 years at a temperature of 85°C. The inductive energy storage element can be integrated into the micro inverter either at DC side or AC side by considering advantages and disadvantages of each method. Such a micro inverter with inductive energy storage-based APD can be found in [49].

4.7   COMPARISON OF POWER DECOUPLING TECHNIQUES

Power decoupling methods that can be used in single-phase micro inverters were discussed in the above sections. The size of the power decoupling capacitor is a major concern in PV applications and it can be reduced using control techniques and additional power converter topologies. Additional power converter topologies added into the basic power converter makes its overall power conversion efficiency to be reduced. The high-voltage DC-link based micro inverters can be used to decouple double-line frequency power ripple without an extra power converter. Size of the power decoupling capacitor can be further reduced using complex power modulation methods such as bandwidth restricted controllers and filters. AC side power decoupling is another feasible option without additional power converters which requires only a small capacitor and that can be fulfilled using a reliable capacitor technology such as film capacitors.

A micro inverter with a third ripple port shown in Fig. 4.11 can be analyzed to get a better understanding about power decoupling operation. The auxiliary power port is magnetically integrated to the main power circuit using an extra winding in the HF isolation transformer. The front-end converter and half-wave cycloconverter are controlled using fixed switching frequency which is above resonant frequency of the series resonant circuit and power is modulated using two phase shifts, namely, phase shift between control signals of two legs of the full-bridge converter and phase shift between HFL resonant current and half-wave cycloconverter input voltage. PWM is used at the auxiliary port to store and release additional power generated by PV module compared to the power demand of the load. Fixed frequency pulse width modulated control signals of the auxiliary power port are generated by comparing the sinusoidal reference signal against a carrier waveform.

The micro inverter output power with and without auxiliary power port is shown in Fig. 4.12. The inclusion of extra power port does not have a significant effect on the output of the micro inverter. However, a significant difference can be observed in a micro inverter input current and voltage with and without auxiliary power port, as shown in Figs. 4.13 and 4.14.

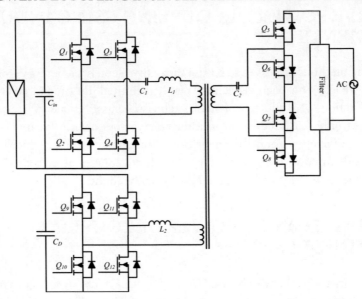

Figure 4.11: A HFL micro inverter with an auxiliary power port to integrate power decoupling capacitor.

A significant reduction can be observed in micro inverter input current in Fig. 4.13 with the introduction of the auxiliary power port. A similar behavior can be observed in the voltage across the micro inverter input buffer capacitor, as shown in Fig. 4.14. Hence, there is a significant reduction in amplitude of the double-line-frequency power ripple at the input port of the micro inverter. That reduction is obtained by storing and releasing energy in the power decoupling capacitor integrated using the auxiliary power port. The voltage variation across the power decoupling capacitor is shown in Fig. 4.15 where a DC quantity is superimposed with the double-line frequency AC ripple.

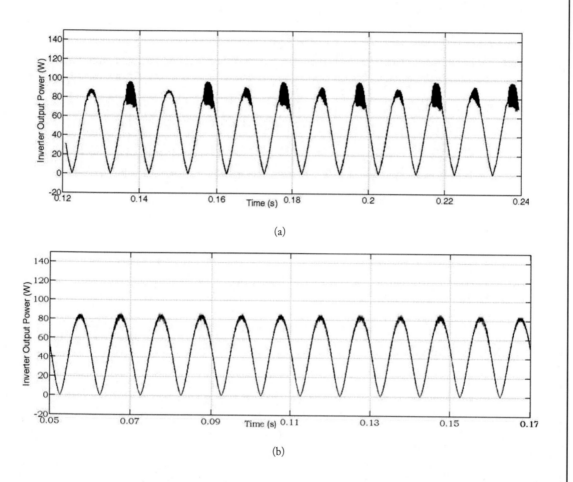

Figure 4.12: Micro inverter output power (a) without and (b) with auxiliary power port.

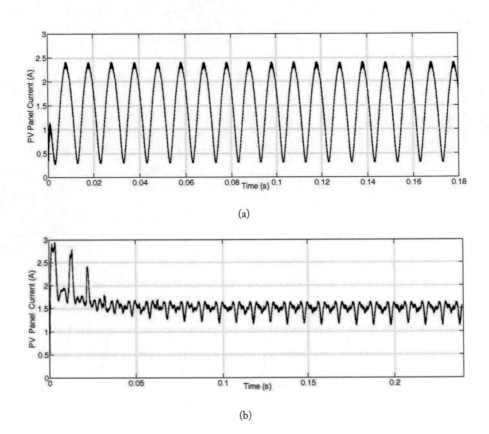

Figure 4.13: Micro inverter input current (a) without and (b) with auxiliary power port.

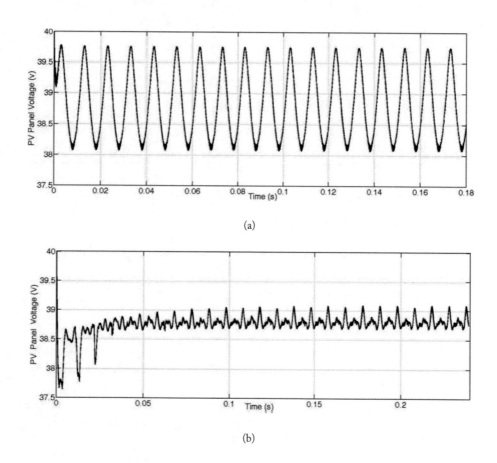

(a)

(b)

Figure 4.14: Terminal voltage of input buffer capacitor (C_{buf}) (a) without, and (b) with auxiliary power port.

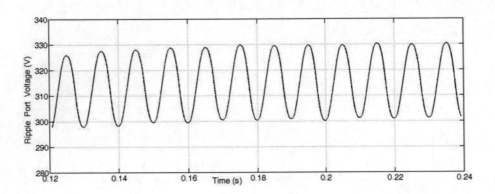

Figure 4.15: Terminal voltage across the power decoupling capacitor at the auxiliary power port.

CHAPTER 5

Energy Storage Interfacing

5.1 INTRODUCTION

Solar irradiance varies with the time of the day and the presence of clouds. As a result, electric power generated in photovoltaic cells varies with time. Moreover, the generated power varies due to change in the incidence angle of light and shading. On the other hand, depending on the system which the PV system is connected, the power demand varies with the time. As a result, at any given time there is a possibility of having a surplus of power or a lack of power in the PV system. But the fundamental requirement, i.e., the balance between generated and the delivered power should be maintained all the time. In off-grid residential PV systems, the PV system is the only supply and therefore there should be an energy storage mechanism that can absorb power fluctuations and thereby maintain the balance. If energy storage is not available, the imbalance appears as fluctuations in the output voltage which might be detrimental for the connected equipment. In grid-connected residential PV systems, the generated PV power is directly fed into the grid. In other words, power fluctuations are transferred into the grid. Similarly, in commercial PV systems it is possible to transfer power fluctuations into the grid if the grid has sufficient amount of spinning reserve to absorb the fluctuations. If the grid does not have sufficient spinning reserve these fluctuations can cause system instabilities. In such situations grid operators recommend adding energy storage to absorb fluctuations.

Utility scale PV systems supply a significant share of power to the grid. Therefore, fluctuations in the generated PV power should be absorbed on-site with suitable energy storage mechanisms and the output should be regulated depending on the grid requirements. Moreover, utility scale large photovoltaic power systems are expected to regulate their power dispatch at least for a one-hour period. Apart from the energy storage, solar power prediction is emerging as a supporting mechanism that can help to estimate the amount of power that can be dispatched in the next hour and thereby reduce the size of the energy storage. Therefore, in summary, energy storage technologies help absorb short-term power fluctuations ranging from seconds to hours while solar power prediction helps to optimize the use of energy storage.

Apart from the mitigation of power fluctuations, energy storage systems can play other important roles as well in PV power systems. Load shifting is one such role which refers to the store of energy during low demand periods and discharge during high demand periods. To meet grid codes in some countries, large PV systems are required to remain connected during short term low voltage conditions of the grid. This is known as the low-voltage ride through (LVRT) capability where energy storage systems can make a significant contribution. The amount of power that can be injected into the grid drops during these low voltage conditions which results in an accumulation of energy within the PV system. This energy accumulation reflects as an increase in the voltage in intermediate stages and lead to instabilities and stresses in the power converter system. This instability makes it difficult to remain connected during the recovery period. If energy storage systems are present, they can be used to store excess energy and thereby keep the system stable and connected during the recovery period. These supporting functions together with the need for absorbing power fluctuations consolidate that energy storage is an inevitable part in modern PV power systems.

Batteries, supercapacitors, flywheels, compressed air, pumped hydro, superconductors, and generation of hydrogen are possible energy storage technologies that can be used in PV systems. Out of these energy storage technologies, batteries and supercapacitors have become the popular choice owing to their wide availability, simplicity, technological maturity, and lower cost compared to the other options. Moreover, the combination of battery and supercapacitor makes a good synergy and helps extend power and energy capacities that cannot be achieved with individual storage systems.

Characteristics of batteries and supercapacitors are different from each other. Moreover, photovoltaic systems have their own characteristics. Therefore, interfacing technologies that can match those characteristics have become an essential part in modern photovoltaic power systems. Power electronic converter systems are the only choice that can meet all these requirements. Therefore, the aim of this chapter is to discuss these characteristics in detail, identify suitable power electronic converter technologies that can be used as interfaces and discuss their advantages and limitations. Recent developments in the interfacing converter technologies are also provided in the later part of this chapter.

5.2   CHARACTERISTICS OF BATTERIES, SUPERCAPCITORS, AND PV CELLS

Lead acid is the most common and widely available battery technology that can be used as energy storage in residential and commercial PV power systems. Even though lead acid batteries are not expensive, their efficiency and cycle life are low compared to the alternative battery technologies such as Lithium-ion, molten-salt and flow batteries. Out of these technologies, Lithium ion batteries have the highest efficiency and cycle life. Therefore, they have become the number one choice in consumer electronic and automobile industries. However, cost of Lithium-ion batteries are still at the high side and therefore not a very popular choice in PV systems. This has created a technology gap in energy storage especially for commercial and utility scale PV power systems. On the other hand, flow battery technologies such as Zinc-Bromide, Vanadium-Redox, and Iron-Chromium and molten-salt battery technologies such as Sodium-Sulfur and Sodium-Nickel are moderate in cost, efficiency, and cycle life. Therefore, these technologies are becoming the popular choice for energy storage in commercial and utility-scale PV power systems.

An equivalent circuit that can be used to explain the behavior of batteries is shown in Fig. 5.1a. Depending on the battery technology there might be slight differences in the respective equivalent circuit and its parameters. Nevertheless, in most cases this circuit is detailed enough to explain battery characteristics. The voltage, V_{SOC}, marked in this equivalent circuit represents the open circuit voltage of the battery. This voltage is a function of the state of charge (SoC) of the battery. However, this voltage usually shows only a little change as shown in Fig. 5.1b. In other words, batteries can supply steady power for a long period of time without a significant drop in the voltage. This explains the capability of batteries to act as an energy source. The resistor in the equivalent circuit represents the internal resistance of the battery. The voltage drop across this resistor reduces the output voltage of the battery when it is discharging. In the same way, this drop has to be compensated during charging and therefore the supply voltage has to be larger than V_{SOC}. This voltage drop is the reason why there are two separate lines in Fig. 5.1b representing the battery charging and discharging. Moreover, this resistance limits the output power and charging power. Therefore, even though batteries can sustain steady power for a long period of time the maximum power that can be delivered or absorbed by the battery is limited by the internal resistance. The other two RC networks represent slow response of batteries to transients. Owing to the high time constants of these RC components, batteries are not capable of supporting fast charging and discharging cycles. Moreover, fast charging/discharging reduces battery life. Therefore, batteries can be used as an energy source but not as a power source for absorbing short-term large power fluctuations.

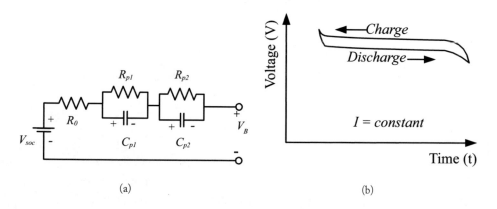

(a)   (b)

Figure 5.1: (a) Battery equivalent circuit and (b) variation of the terminal voltage under constant current charging and discharging.

An equivalent circuit of a supercapacitor is shown in Fig. 5.2a. Both the capacitances and the resistances shown in this figure vary with the frequency, temperature, and voltage. But for simplicity, these large number of RC branches can be simplified into a simple RLC network, as shown Fig. 5.2b. In this figure, the effect of resistors are simplified into an equivalent series

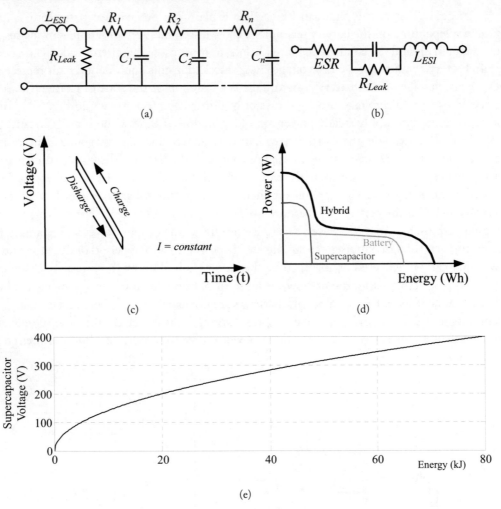

Figure 5.2: (a) Detailed equivalent circuit of a supercapacitor, (b) simplified equivalent circuit of a supercapacitor, (c) variation of the supercapacitor voltage against time for constant current charging and discharging, and (d) change of supercapacitor voltage with stored energy.

resistance (ESR) and a leakage resistance R_{Leak}. Similar to the batteries, ESR introduces voltage drop and limits the charging/discharging power. However, this resistance is very low compared to that of the batteries. As a result, supercapacitors are capable of supplying or absorbing significantly large amount of power compared to batteries. This makes supercapacitors an ideal choice for improving the LVRT capability. As the supercapacitor gets discharged its voltage drops and therefore, a high level of output power cannot be maintained for a long time, as shown in Fig. 5.2b.

Therefore, supercapacitors are good only as a high power source but not as an energy source. As illustrated in Fig. 5.2d, the combination of battery and supercapacitor can make a good match where supercapacitors help mitigate short term power fluctuations and batteries help mitigate long-term power fluctuations. Furthermore, this combination reduces fast charging/discharging stresses on the battery and thereby prolong the battery life. The leakage resistance is quite low in supercapacitors and therefore their self-discharge rate is high compared to the batteries.

Energy stored in a capacitor has a square relationship with the voltage as expressed in (5.1) where E is the stored energy, C is the capacitance, and V is the voltage. The corresponding change of the voltage with the stored energy is shown in Fig. 5.2e. This diagram further indicates that more than 75% of energy can be taken out of the supercapacitor with a 50% drop in its voltage.

$$E = \frac{1}{2}CV^2. \tag{5.1}$$

5.3   THE NEED OF ENERGY STORAGE INTERFACING IN PV SYSTEMS

A typical voltage-current characteristic of a photovoltaic array is shown in Fig. 5.3a. As this diagram illustrates, terminal voltage of PV cells depend on the solar insolation and the current drawn from the cell. As a result, the output power becomes maximum only at a particular point and it varies with the solar insolation. The change of this maximum power point with solar irradiance shows that the output voltage at the maximum power point increases with the solar insolation. Therefore, if maximum power point tracking (MPPT) is to be implemented, the terminal voltage of the PV should be given freedom to vary within a range marked as PV panel output voltage.

The simplest way of integrating a battery with the PV panel is the direct connection across the panel as shown in Fig. 5.4a. However, when a battery is in parallel with the PV panel, it tries to control the terminal voltage of PV module and as a result the maximum power point tracking becomes impossible. In other words, the power drawn from the PV panel depends on the battery voltage. For example, if the dashed vertical line in Fig. 5.3a corresponds to the battery voltage at any given instance, the possible PV power levels are the points that intersect with this line. These are not the maximum power points in most of the cases. Therefore, even though this direct connection is simple, it reduces the efficiency of PV power capture. Moreover, this arrangement does not allow controlling the battery power and therefore fluctuations present in PV power is directly passed to the battery. The situation becomes even worse if a supercapacitor is directly connected across the PV panel as its voltage varies in a wide range. Therefore, in order to achieve the maximum power capture, PV panel should be given the freedom to vary its output voltage within the optimum range. Energy storage elements such as batteries and supercapacitors need to be segregated from the PV panel. The most common way of making this separation is the use of a DC-DC converter, as shown in Fig. 5.4b.

The DC-DC boost converter in Fig. 5.4b acts as an interface between the PV panel and the grid connecting inverter. This allows the PV panel to vary its output voltage within the optimum

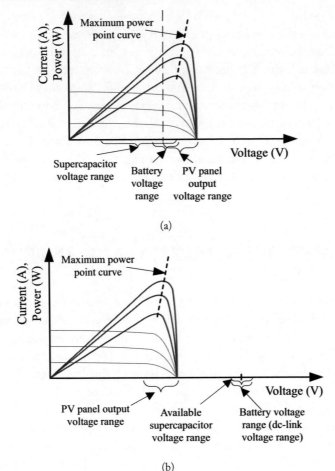

Figure 5.3: PV characteristic and voltage ranges of battery and supercapacitor (a) direct connection of energy storage element(s) across the PV panel and (b) direct connection of energy storage element(s) to the DC-link.

range and thereby operate at the maximum power point (MPP). The output voltage of the DC-DC boost converter, known as the DC-link voltage, V_{dc}, is generally regulated to a level higher than the output voltage of the PV panel, as shown in Fig. 5.3b. However, there is a possibility of letting the DC-link voltage vary in a small range a so that a battery can be connected directly into the DC-link as shown in Fig. 5.4b. This voltage range is marked in Fig. 5.3b as the battery voltage range. Unlike in the previous case, this configuration allows the control of the battery power with the appropriate control of the boost converter. The two battery interfacing techniques shown in

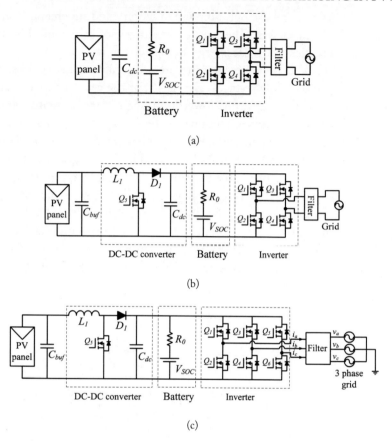

Figure 5.4: (a) A battery is directly connected across the PV panel in a single-phase system, (b) interfacing DC-DC converter for the PV panel and direct connection of a battery to the DC-link in a single-phase system, and (c) interfacing DC-DC converter for the PV panel and direct connection of a battery to the DC-link in a three-phase system.

Figs. 5.4a and 5.4b are for feeding a single-phase power grid. The hardware implementation for three-phase power systems is an extension of the single-phase system, as shown in Fig. 5.4c.

Even though the direct connection of the battery to the DC-link is simple, it allows the PV panel to track the MPP and control battery power so that the fluctuation present in the PV power is directly passed to the battery. Moreover, as the DC-link voltage gets higher, more battery cells are required to be connected in series to meet the voltage requirement. This increases the internal resistance and cell voltage balancing becomes more complex. Apart from that, due to the wide voltage range requirement for supercapacitors to get their maximum use, it is not possible to connect them directly into the DC-link as the allowable variation in the DC-link voltage is limited. Therefore, in order to get the optimum use of battery and supercapacitor, they should be segregated from the DC-link as well. The most common way of achieving this is to use separate interfacing DC-DC converters for these energy storage elements.

The interfacing DC-DC converters need to provide a high voltage gain so that the energy storage elements can operate at low voltages. Especially, in the case of supercapacitors, the DC-DC converter needs to operate even at the lower limit of the supercapacitor voltage. Moreover, the interfacing converters should be bi-directional as the energy storage elements are supposed to have both charging and discharging functionalities. In other words, they should be able to operate in first two quadrants as shown in Fig. 5.5. As illustrated in this diagram, both input and output voltages of the converters are positive while the current flow can be either in positive or negative direction. Battery and supercapacitor voltages are generally kept lower than the DC-link voltage. When the energy storage feeds power to the DC-link, the interfacing DC-DC converter needs to act as a boost converter to facilitate the battery and/or supercapacitor discharging process. On the other hand, when the energy storage system is required to draw power from the DC-link, the interfacing DC-DC converter needs to act as a buck converter to facilitate the battery and/or supercapacitor charging process.

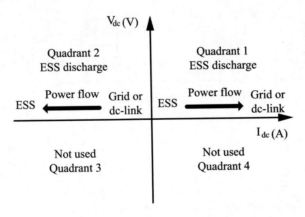

Figure 5.5: DC-DC converter operating areas in the four-quadrant diagram.

5.4   COMMONLY USED ENERGY STORAGE INTERFACING CONVERTER TOPOLOGIES

The simplest bidirectional DC-DC converter that can be used to interface a battery to the DC-link of the PV power conversion system is shown in Fig. 5.6a. This converter can be considered as a combination of traditional buck and boost converters. As this converter uses only two switches with anti-parallel diodes and an inductor at the low voltage side, it is simple to control and cost effective to implement. The identical converter topology can be used to interface a supercapacitor as well to the DC-link as shown in Fig. 5.6b. With the proper control of the switches Q_6-Q_9, it is possible to direct short-term power fluctuations towards the supercapacitor and long-term power fluctuations to the battery.

As in the traditional DC-DC converter, the operation of the bi-directional DC-DC converter shown in Fig. 5.6a is based on charging and discharging of the inductor. Based on inductor charging/discharging and converter boost/buck operation, four different operating modes can be defined for this converter. These four operating modes are illustrated in Fig. 5.6c with corresponding current paths and the switch to be turned on to activate the mode. Inductor current variations and transistor switching states are shown in Fig. 5.6d. As shown in the top two diagrams in Fig. 5.6c, during the boost operation, the transistor Q_6 is turned off. When the complementary transistor Q_7 is turned on, the inductor gets charged form the battery as shown in the top left diagram of Fig. 5.6c. The inductor current waveform is shown in Fig. 5.6d. When Q_7 is turned off, the inductor tries to maintain the current flow and as a result the anti-parallel diode of Q_6 gets forward biased. This creates a current path and let the energy stored in the inductor to discharge to the DC-link, as shown in the top right diagram of Fig. 5.6c.

During the buck mode of operation of the converter, transistor Q_7, is turned off and Q_6 is turned on and off to control inductor charging and discharging as illustrated in the two bottom diagrams of Fig. 5.6c. When Q_6 is turned on, the inductor gets charged as shown in the bottom right diagram of Fig. 5.6c. In this mode of operation, the inductor current direction reverses as shown in Fig. 5.6e. When Q_6 is turned off, the inductor tries to maintain the current flow and as a result the anti-parallel diode of Q_7 gets forward biased. This creates a current path and let the energy stored in the inductor to discharge to the battery as shown in the bottom left diagram of Fig. 5.6c.

Even though the DC-DC converter shown in Fig. 5.6a is simple and cost effective it has certain disadvantages such as ripples in the current, lower voltage transfer gain and the need for a large inductor to reduce current ripple. Generally, it is preferred to charge/discharge batteries with low ripple current in order to achieve higher efficiency and a longer lifetime. Therefore, as a solution to the current ripple and to reduce the inductor size, another DC-DC converter can be connected in parallel as shown in Fig. 5.7a. With this configuration, current ripples can be reduced if the turn on and off operations of the converters are offset by a half-cycle period. This is known as the interleaved operation. Moreover, as the two converters share the power, it is possible to reduce the inductor size of each converter and current ratings of the transistors and diodes. Moreover, it is possible to add more converters in parallel to increase the current rating of the total converter system and thereby reduce the ripple in the current.

Even though the interleaved DC-DC converter reduces current ripples and inductor size, the voltage gain remains the same. Therefore, if a high voltage gain is required then the cascade connection shown in Fig. 5.7b has to be used. Even though it is possible to connect more converters in cascade to achieve higher gains it complicates the converter control. Moreover, this topology requires additional DC capacitors to be placed in between converters.

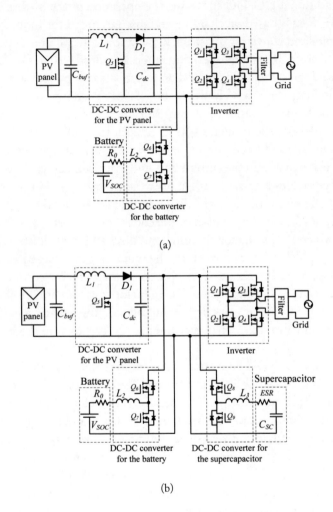

(a)

(b)

Figure 5.6: The simplest bi-directional DC-DC converter for energy storage interfacing (a) a battery is connected to the DC-link through the converter and (b) a battery and supercapacitor are connected to the DC-link through separate DC-DC converters. *(Continues.)*

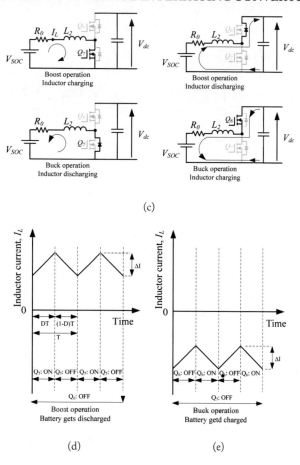

(c)

(d)   (e)

Figure 5.6: *(Continued.)* The simplest bi-directional DC-DC converter for energy storage interfacing, (c) current flow in the four operating modes, (d) inductor current in the boost operation, and (e) inductor current in the buck operation.

The bi-directional DC-DC converter discussed above can only operate in buck mode when charging the battery and boost mode when discharging the battery. This is not an issue as the battery voltage is usually kept below the DC-link voltage. However, if the voltage of the energy storage system is close to the DC-link voltage this converter won't work. For such systems, the buck-boost or Cúk converters are good choices. Figure 5.8a shows a buck-boost converter. The polarity of output voltage gets reversed with respect to a common ground in the buck-boost converter. This might be a burden in certain applications. This issue can be solved by adding more switches as shown in Fig. 5.8b. This configuration is known as the cascade buck-boost converter.

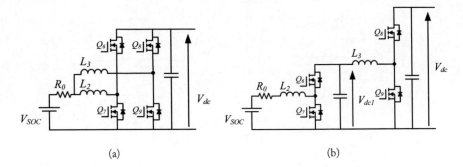

Figure 5.7: (a) Interleaved DC-DC converter and (b) series connection of two DC-DC converters.

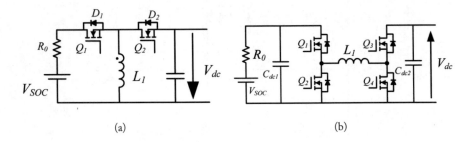

Figure 5.8: (a) Buck-boost type bi-directional DC-DC converter, and (b) cascade bi-directional buck-boost converter.

A few other bi-directional DC-DC converter topologies are shown in Fig. 5.9. Figures 5.9a and 5.9b are two different forms of the Cúk converter. Figure 5.9c is a bidirectional form of a positive output Luo converter. All these converters consist of two switches, two diodes, and two inductors. The main advantage of these topologies is the reduced input and output current ripples. However, they require two large inductors and capacitors which are bulky and expensive. Therefore, these topologies are mostly used in low power applications.

The DC-DC converters discussed so far do not possess any galvanic isolation which is a requirement mandated by many standards for PV systems. Main reasons for the need of galvanic isolation are safety, proper operation of protection systems, and noise reduction. Moreover, in many cases energy storage systems are preferred to be kept at a lower voltage than that of the DC-link and that requires very high gain DC-DC converters. Therefore, if both galvanic isolation and high voltage transfer gain are required, magnetic coupling with a high frequency transformer can be chosen as shown in Fig. 5.10. The high frequency operation reduces the size of the transformer. A high voltage transfer gain can be obtained with a high turns ratio, n, of the transformer.

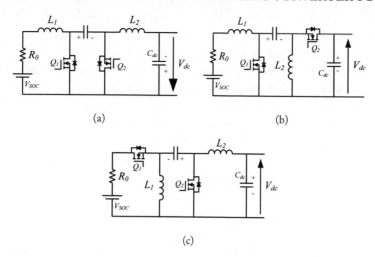

Figure 5.9: (a, b) Bi-directional forms of Cúk converter, and (c) Luo DC-DC converter.

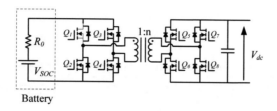

Figure 5.10: Bi-directional isolated full bridge DC-DC converter.

The bidirectional operation of the isolated full-bridge converter shown in Fig. 5.10 requires both converters on either side of the transformer to be equipped with controllable switches. However, only the source-side converter control is sufficient to achieve the power transfer. The diodes on the other side can passively connect AC power to DC power. In other words, if the battery is to be discharged, control on the battery side switches is sufficient to ensure desired power flow from the battery to the DC-link. The switches on the DC-link side can be kept turned off. In the DC-link side, the current flows through the diodes. Similarly, when the battery is to be charged, control on the DC-link side switches is sufficient to achieve the desired power flow. Even though this one side control is simple and reduces control complexity, it does not allow the full use of capabilities and features of the converter system. Therefore, the trend is to control both sides irrespective of the direction of power flow.

In commercial and utility-scale PV power systems, it is economical to have a centralized large energy storage system rather than having separate energy storage systems attached to individual converters such as micro inverters or micro converters. In such applications, the connection of energy storage systems to the AC bus through inverters is preferred. Two possible ways of connecting a battery-supercapacitor hybrid energy storage system are shown in Figs. 5.11a and 5.11b. In both approaches the battery is directly connected to the DC-side of the grid connecting inverter. This direct connection is possible due to the fact that the battery voltage varies only within a small range. However, it is not possible to connect a supercapacitor directly into the DC-side of the inverter as its voltage varies in a wide range. Therefore, an interfacing DC-DC converter is required to connect a supercapacitor to the grid-connecting inverter.

As shown in Fig. 5.11a, it is possible to have separate inverters for the battery and supercapacitor. This gives more control flexibility as their power can be controlled separately. However, the drawback of this approach is the increased component count. The approach shown in Fig. 5.11b is an alternative which slightly compromises the control flexibility over the component count. Even with this approach it is possible to direct short term power fluctuations to the supercapacitor and long term power fluctuations to the battery. Proper control of the switches Q_5–Q_{10} is essential to achieve this objective. As discussed earlier, if the battery voltage is required to be kept at a low value, an additional DC-DC converter needs to be used to interface the battery with the DC-link of the grid connecting inverter. Moreover, the galvanic isolation can be achieved with the use of line frequency transformer at the grid side or isolated DC-DC converters at the DC-side of the grid-connecting inverter.

5.5   SOFT-SWITCHING BASED-ISOLATED BI-DIRECTIONAL DC-DC CONVERTERS FOR ENERGY STORAGE INTERFACING

As mentioned earlier, galvanic isolation and voltage matching have become major considerations in bi-directional DC-DC converters. The bi-directional full-bridge DC-DC converter with a high frequency isolation transformer, shown in Fig. 5.12, is a good solution that can meet those requirements. However, owing to the increased component count, power losses increase and as a result efficiency drops. Losses in the power conversion occur in two forms namely, conduction loss and switching loss. The conduction loss increases with the number of devices in the conduction path. The switching loss increases with the switching frequency. In high frequency converters switching losses are dominant. Therefore, soft-switching technologies are developed to reduce switching losses and thereby improve the power conversion efficiency. Compared to traditional hard-switched PWM converters, the soft-switching converters might have a high circulating current and more devices in the conduction path lead to increase in conduction losses. However, at high switching frequencies, switching loss reduction caused by soft-switching offsets the increase in conduction losses and hence there is an improvement in the overall efficiency.

Soft-switching techniques can be incorporated into the isolated bi-directional full bridge DC-DC converter with resonant tanks as shown in Fig. 3.12. In the series resonant converter shown in Fig. 5.12a the resonant inductor L_r and resonant capacitor C_r are connected in series

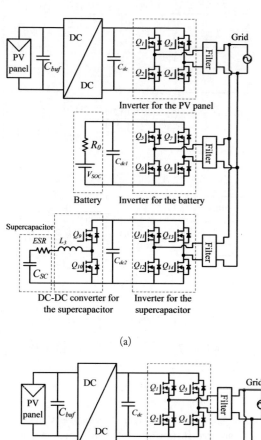

(a)

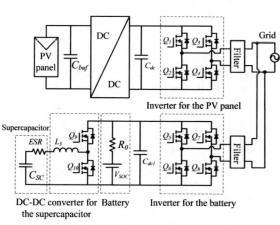

(b)

Figure 5.11: (a) Battery and supercapacitor connected to the AC bus through DC-AC inverters, and (b) alternative way of integrating a battery and a supercapacitor to the AC bus through a DC-AC inverter.

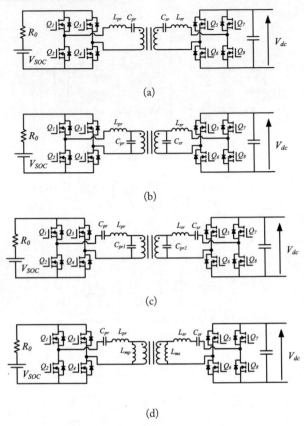

Figure 5.12: Bi-directional isolated full bridge DC-DC converter with (a) series resonance, (b) parallel resonance, (c) series parallel resonance, and (d) LLC resonance.

with active full-bridge switches in both sides of the transformer. These inductors and capacitors form resonance tanks in both sides of the transformer and act as voltage dividers with the load reflected from the other side of the transformer. At the resonant frequency, the impedance of the series resonant tank is insignificant and therefore, the resonance does not reduce the voltage gain. If the switching frequency is slightly increased or decreased from the resonant frequency the impedance of the resonant tank increases rapidly and therefore, the voltage gain becomes lower. Therefore, this converter has a limited operating range. In order to increase the efficiency, zero voltage switching (ZVS) is preferred if the switching frequency is larger than the resonance frequency and zero current switching (ZCS) is preferred when the switching frequency is lower than the resonance frequency [50]. The main drawbacks of the series resonance topology are the need for high switching frequency to regulate output voltage and high circulating energy at light load conditions.

The parallel resonance converter shown in Fig. 5.12b has a wider operating range compared to the series resonance converter. However, high circulating energy can be seen at light load conditions. The series parallel resonance converter shown in Fig. 5.12c merges the above two resonant topologies with two resonance capacitors and one resonance inductor. This combination helps minimize limitation of each resonant topology and therefore it is possible to achieve voltage regulation even at no-load conditions. The main drawback of the series parallel resonance converter is drop in voltage transfer gain if the voltage range is large. The LLC topology shown in Fig. 5.12d is an alternative to achieve high switching frequency and higher efficiency at high voltages [50].

5.6   SIMULATION STUDY

This section is aimed at explaining the control aspect of energy storage interfacing converters and relevant results are obtained in a simulation study. As shown in Fig. 5.13a, a PV power system feeding a three-phase power grid has been selected for this study. The battery bank is connected to the DC-link through an interfacing DC-DC converter. The control of the entire system can be delegated into three distinct controllers, namely, MPPT controller, DC-DC converter controller, and the grid connecting inverter controller which are shown in Figs. 5.13b, 5.13c, and 5.13d, respectively.

The MPPT control system shown in Fig. 5.13a consists of three main components, namely: the MPP algorithm, outer voltage controller, and inner current controller. The MPP algorithm ensures maximum power is extracted from the PV panel for a given insolation level. In order to do this, the MPP algorithm calculates the optimal PV panel terminal voltage using the power gradient algorithm known as dP/dV control. At the MPP, the power gradient, dP/dV, should be zero. If it is positive that means the optimal point has not been reached and it is possible to extract more power. Therefore, the PV panel terminal voltage has to be increased further. Similarly, a negative gradient indicates that the increase of the PV terminal voltage will decrease the power that can be extracted from the PV panel. Therefore, the terminal voltage should be reduced. Based on the decision of the MPPT algorithm, a reference voltage will be generated which feeds the subsequent outer voltage controller loop that in turn regulates the PV panel terminal voltage to this optimal level. This control is achieved by controlling the innermost current loop. The output of the innermost current controller is the modulation index that is sent to the PWM module. The PWM module generates required gate signals for the switching device.

The control objective of the battery interfacing DC-DC converter is to regulate the DC-link voltage to a predefined value. This is done by either injecting power into the DC-link or absorbing power from that. The V^2 controller shown in Fig. 5.13c generates a positive modulation index depending on the amount of power exchange required between the battery and the DC-link. If the modulation index is positive, that means the battery should inject power to the DC-link. This is done by activating the boost converter. Similarly, if the modulation index is negative, that

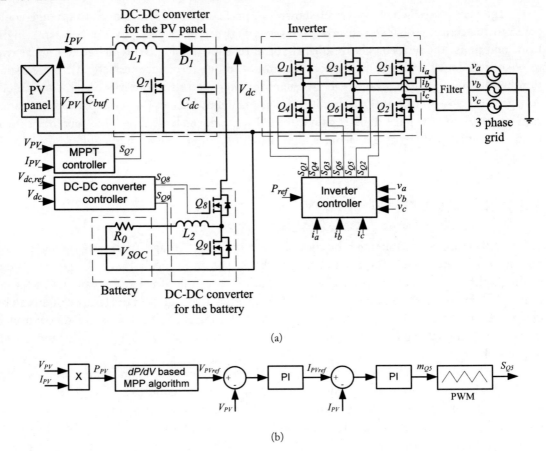

(a)

(b)

Figure 5.13: (a) Schematic of the PV power system used in the simulation study and (b) MPPT controller. *(Continues.)*

means power should be absorbed by the DC-link. This can be achieved by activating the buck converter. The corresponding switching signal generation logic is shown in Fig. 5.13c.

The grid connecting inverter regulates the current fed into the grid. The amounts of active and reactive power transferred to the grid are determined by the d- and q-axes current references, i_d and i_q, shown in Fig. 5.13d. Generally, the q-axis current, i_q, which determines the reactive power exchange, is set to zero to achieve unity power factor. Therefore, in this simulation study, the q-axis current reference, i_{qref}, is set to zero. The d-axis current, i_d, which determines the active power exchange with the grid, is controlled by an outer control loop which is not shown in this diagram. In this simulation, the d-axis current reference, i_{dref}, is set to a constant value which represents a constant power dispatch situation. These two current references are compared

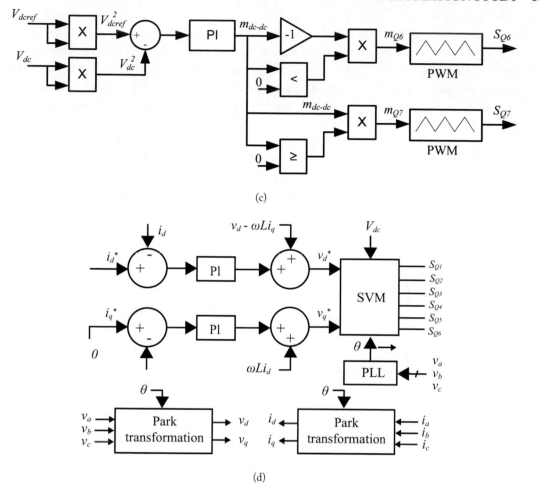

(c)

(d)

Figure 5.13: *(Continued.)* (c) DC-DC converter controller, and (d) grid interfacing inverter controller.

with the actual currents to obtain errors which are then passed through PI controllers as shown in Fig. 5.13d to obtain voltage references, v_{dref}, and v_{qref}. The space vector modulation (SVM) has been used to generate gate signals for the switches. The orientation reference θ for the modulator and dq axis transformation of voltage and currents is provided by a phase-locked loop (PLL). The Park transformation is used in this dq axis transformation.

In the simulation, two step changes are introduced to the solar insolation, as shown in Fig. 5.14a, aiming to illustrate the dynamic behavior of the aforementioned three power converters of the PV system. As shown in Fig. 5.14b, power generated by the PV panel varies with the change in the solar insolation. The battery energy storage system absorbs these fluctuations and thereby maintains a constant power dispatch to the grid as shown by the trace marked as P_g in Fig. 5.14b. This is achieved by charging the battery when the PV power is high and discharging the battery when it is low. In other words, battery gets charged when the solar insolation is at 0.9 and discharged when it drops to 0.2. The captured PV power is slightly above the output power dispatched to the grid when the solar insolation is at 0.6 and therefore the battery still gets charged but at a lower power level. The corresponding battery power variations are depicted in Fig. 5.14b by the trace marked as P_b.

The corresponding variations in the PV panel terminal voltage, battery terminal voltage and DC-link voltage are shown in Fig. 5.14c. As mentioned above, with reference to Fig. 5.3, the PV panel terminal voltage shows a slight drop when the solar insolation drops from 0.9 to 0.2. This is a result of the maximum power point tracking achieved by the controller of the DC-

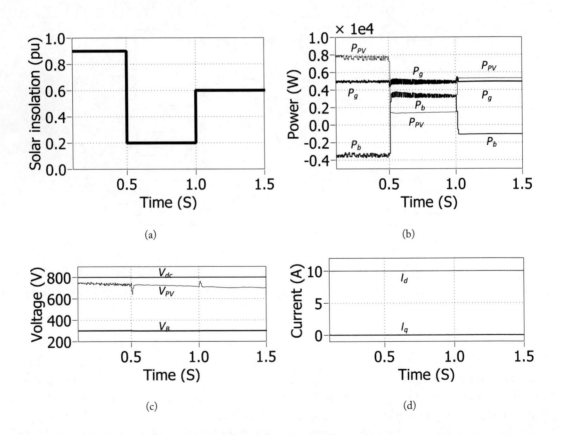

Figure 5.14: (a) Solar insolation, (b) PV power, battery power and power dispatched to the grid, (c) PV voltage, battery voltage and DC-link voltage, and (d) dq-axis component of the inverter output current. *(Continues.)*

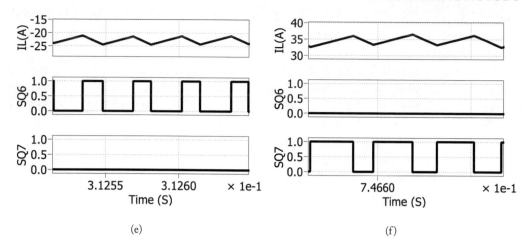

(e)   (f)

Figure 5.14: *(Continued.)* (e) inductor current and gate signals of the switches S_6 and S_7 during the boost operation, and (f) inductor current and gate signals of the switches S_6 and S_7 during the boost operation.

DC converter attached to the PV panel. The DC-link voltage is regulated at the set point by the DC-DC converter. The battery terminal voltage shows a slight increase when it is charging compared to discharging. This accounts for the voltage drop across the internal resistance of the battery bank.

The control objective for the grid connecting inverter is to maintain a constant power delivery to the grid irrespective of changes in the captured PV power. This constant power delivery to the grid is achieved by regulating the d-axis current component, i_d, of the inverter output current. In this simulation, the d-axis current reference is set to 10A which corresponds to 5 kW power output to the grid. The q-axis current component, i_q, is regulated at zero to ensure unity power factor at the feeding point. The corresponding d-axis and q-axis currents are shown in Fig. 5.14d which are well regulated at the set points irrespective of the changes present in the captured solar power.

In order to illustrate the operation of the battery interfacing bi-directional DC-DC converter during charging and discharging states, inductor current variations and gate signals of the two switches are shown in Figs. 5.14e and 5.14f, respectively. As discussed above with reference to Figs. 5.6d and 5.6e when the battery is charging the converter operates as a buck converter. The switch Q_7 is turned off as illustrated in SQ_7 plot in Fig. 5.14f. The switch Q_6 turns on and off to let the inductor gets charged and discharged. As depicted in the plot of I_L in Fig. 5.14f, when the switch Q_6 is turned on, the inductor gets charged. According to the convention used in Fig. 5.13a, the current flowing out of the battery is positive. Therefore, in this buck converter operation, the inductor current is negative and increases with a negative slope. Similarly, when the switch Q_6 is turned off the inductor gets discharged with a positive slope in the current wave-form. As the inductor current is negative during the buck converter operation this positive slope indicates a decrease of the current.

An identical analysis can be extended for the boost mode of operation where the battery gets discharged to supply power to the DC-link. The corresponding inductor current variations and gate signals of the two switches are shown in Fig. 5.14f. As shown in SQ_6 and SQ_7 plots of Fig. 5.14f, in the boost mode of operation, switch Q_6 is kept turned-off and Q_7 is turned on and off to charge and discharge the inductor. The inductor current is positive in the boost mode of operation. When switch Q_7 is turned on, the inductor current increases with a positive slope which indicates that it gets charged during this period. When switch Q_7 is turned off the inductor discharges its stored energy into the DC-link.

References

[1] M. Dale and S.M. Benson, "Energy balance of the global photovoltaic (PV) industry - is the PV industry a net electricity producer?," *Environ. Sci. Technol.*, vol. 47, no. 7, pp. 3482–3489, 2013. DOI: 10.1021/es3038824. 1

[2] V. Fthenakis, "Sustainability of photovoltaics: The case for thin-film solar cells," *Renew. Sustain. Energy Rev.*, vol. 13, pp. 2746–2750, 2009. DOI: 10.1016/j.rser.2009.05.001. 2

[3] F. Kessler, D. Herrmann, and M. Powalla, "Approaches to flexible CIGS thin-film solar cells," *Thin Solid Films*, vol. 480–481, pp. 491–498, 2005. DOI: 10.1016/j.tsf.2004.11.063. 2

[4] N. Espinosa, R. García-Valverde, A. Urbina, and F.C. Krebs, "A life cycle analysis of polymer solar cell modules prepared using roll-to-roll methods under ambient conditions," *Sol. Energy Mater. Sol. Cells*, vol. 95, pp. 1293–1302, 2011. DOI: 10.1016/j.solmat.2010.08.020. 2

[5] H. Tributsch, "Nanocomposite solar cells: The requirement and challenge of kinetic charge separation," *J. Solid State Electrochem.*, vol. 13, pp. 1127–1140, 2009. DOI: 10.1007/s10008-008-0668-2. 2

[6] S. Guha, J. Yang, and B. Yan, "High efficiency multi-junction thin film silicon cells incorporating nanocrystalline silicon," *Sol. Energy Mater. Sol. Cells*, vol. 119, pp. 1–11, 2013. DOI: 10.1016/j.solmat.2013.03.036. 2

[7] M. Elborg, T. Noda, T. Mano, M. Jo, Y. Sakuma, K. Sakoda, and L. Han, "Solar energy materials & solar cells voltage dependence of two-step photocurrent generation in quantum dot intermediate band solar cells," *Sol. Energy Mater. Sol. Cells*, vol. 134, pp. 108–113, 2015. DOI: 10.1016/j.solmat.2014.11.038. 2

[8] M.G. Villalva, J.R. Gazoli, and E.R. Filho, "Comprehensive approach to modeling and simulation of photovoltaic arrays," *IEEE Trans. Power Electron.*, vol. 24, no. 5, pp. 1198–1208, May 2009. DOI: 10.1109/TPEL.2009.2013862. 8

[9] E. Lorenzo, *Solar electricity: Engineering of photovoltaic systems*. Progensa. 9

[10] Y. Xue, K.C. Divya, G. Griepentrog, M. Liviu, S. Suresh, and M. Manjrekar, "Towards next generation photovoltaic inverters," in *Proc. IEEE ECCE*, 2011, pp. 2467–2474. DOI: 10.1109/ECCE.2011.6064096. 15

[11] M. Liserre, T. Sauter, and J.Y. Hung, "Future energy systems: integrating renewable energy sources into the smart power grid through industrial electronics," *IEEE Ind. Electron. Mag.*, vol. 4, no. March, pp. 18–37, 2010. DOI: 10.1109/MIE.2010.935861. 17

[12] T. Esram and P.L. Chapman, "Comparison of photovoltaic array maximum power point tracking techniques," *IEEE Trans. Energy Convers.*, vol. 22, no. 2, pp. 439–449, 2007. DOI: 10.1109/TEC.2006.874230. 17

[13] F.Z. Peng, "Z-source inverter," *IEEE Trans. Ind. Appl.*, vol. 39, no. 2, pp. 504–510, 2003. DOI: 10.1109/TIA.2003.808920. 25

[14] M. Victor, F. Greizer, S. Bremicker, and H. Uwe, "Method of converting a direct current voltage from a source of direct current voltage, more specifically from a photovoltaic couse of direct current voltage, into a alternating current voltage," US 2005/0286281 A12005. 30

[15] R. González, J. López, P. Sanchis, and L. Marroyo, "Transformerless inverter for single-phase photovoltaic systems," *IEEE Trans. Power Electron.*, vol. 22, no. 2, pp. 693–697, 2007. DOI: 10.1109/TPEL.2007.892120. 31

[16] H. Schmidt, C. Siedle, and J. Ketterer, "Current inverter for direct/alternating currents, has direct and alternating connections with an intermediate power store, a bridge circuit, rectifier diodes and a inductive choke," German Patent DE10 221 592 A12003. 31, 32

[17] P. Knaup, "Inverter," US20090003024 A12009. 37

[18] J. Mei, B. Xiao, K. Shen, L.M. Tolbert, and J.Y. Zheng, "Modular multilevel inverter with new modulation method and its application to photovoltaic grid-connected generator," *IEEE Trans. Power Electron.*, vol. 28, no. 11, pp. 5063–5073, 2013. DOI: 10.1109/TPEL.2013.2243758. 39

[19] R. Teodorescu, M. Liserre, and P. Rodríguez, *Grid Converters for Photovoltaic and Wind Power Systems*. John Wiley & Sons, 2010, pp. 5–29. DOI: 10.1002/9780470667057. 41

[20] Enecsys, "Enecsys duo micro inverter," 2013. [Online]. Available: http://www.enecsys.com/products/micro-inverter-duo/. 48

[21] L.M.L. Ma, T. Kerekes, R. Teodorescu, X.J.X. Jin, D. Floricau, and M. Liserre, "The high efficiency transformer-less PV inverter topologies derived from NPC topology," *2009 13th Eur. Conf. Power Electron. Appl.*, 2009. 51

[22] R. West, "Monopolar DC to bipolar DC to AC converter," US 2008/0037305 A12008. 52

[23] J.L. Duran-Gomez, E. Garcia-Cervantes, D.R. Lopez-Flores, P.N. Enjeti, and L. Palma, "Analysis and evaluation of a series-combined connected boost and buck-boost DC-DC converter for photovoltaic application," in *Proc. IEEE APEC*, 2006, pp. 979–985. DOI: 10.1109/APEC.2006.1620657. 52

[24] S.V. Araújo, P. Zacharias, and B. Sahan, "Novel grid-connected non-isolated converters for photovoltaic systems with grounded generator," in *Proc. IEEE PESC*, 2008, pp. 58–65. DOI: 10.1109/PESC.2008.4591897. 52

[25] F.-S. Kang, C.-U. Kim, S.-J. Park, and H.-W. Park, "Interface circuit for photovoltaic system based on buck-boost current-source PWM inverter," in *Proc. IEEE IECON*, 2002, pp. 3257–3261. DOI: 10.1109/IECON.2002.1182920. 54

[26] S. Saha and V.P. Sundarsingh, "Novel grid-connected photovoltaic inverter," *IEE Proc. Gener. Transm. Distrib.*, vol. 143, p. 219, 1996. DOI: 10.1049/ip-gtd:19960054. 54

[27] O. Abdel-rahim, M. Orabi, and M.E. Ahmed, "High gain single-stage inverter for photovoltaic AC modules," in *Proc. IEEE APEC*, 2011, pp. 1961–1967. DOI: 10.1109/APEC.2011.5744865. 54

[28] Z. Zhao, M. Xu, Q. Chen, and Y.C. Jih-Sheng (Jason) Lai, "Derivation, analysis, and implementation of a boost–buck converter-based high-efficiency PV inverter," *IEEE Trans. Power Electron.*, vol. 27, no. 3, pp. 1304–1313, 2012. DOI: 10.1109/TPEL.2011.2163805. 54

[29] L.G. Junior, M.A.G. De Brito, L.P. Sampaio, and C.A. Canesin, "Integrated single-stage converters with tri-state modulation suitable for photovoltaic systems," in *Proc. IEEE ISIE*, 2011, pp. 436–443. DOI: 10.1109/COBEP.2011.6085318. 55

[30] D. Cao, S. Jiang, X. Yu, and F.Z. Peng, "Low-cost semi-Z-source inverter for single-phase photovoltaic systems," *IEEE Trans. Power Electron.*, vol. 26, no. 12, pp. 3514–3523, 2011. DOI: 10.1109/TPEL.2011.2148728. 55

[31] D.C. Martins and R. Demonti, "Photovoltaic energy processing for utility connected system," in *Proc. IEEE IECON*, 2001, vol. 00, no. C, pp. 1965–1969. DOI: 10.1109/IECON.2001.975592. 58

[32] P. Wolfs and Q. Li, "An analysis of a resonant half bridge dual converter operating in continous and discontinous modes," in *Proc. IEEE PESC*, 2002, pp. 1313–1318. DOI: 10.1109/PSEC.2002.1022358. 59

[33] S. Mekhilef, N.A. Rahim, and A.M. Omar, "A new solar energy conversion scheme implemented using grid-tied single phase inverter," in *Proc. IEEE TENCON*, 2000, vol. 3, pp. 524–527. DOI: 10.1109/TENCON.2000.892322. 59

[34] D.C. Martins and R. Demonti, "Interconnection of a photovoltaic panels array to a single-phase utility line from a static conversion system," in *Proc. IEEE PESC*, 2000, vol. 3, pp. 1207–1211. DOI: 10.1109/PESC.2000.880483. 59

[35] C. Prapanavarat, M. Barnes, and N. Jenkins, "Investigation of the performance of a photovoltaic AC module," *IEE Proc.*, vol. 149, no. 4, pp. 472–478, 2002. DOI: 10.1049/ip-gtd:20020141. 59

[36] D.R. Nayanasiri, D.M. Vilathgamuwa, and D.L. Maskell, "Half-wave cycloconverter-based photovoltaic micro inverter topology with phase-shift power modulation," *IEEE Trans. Power Electron.*, vol. 28, no. 6, pp. 2700–2710, 2013. DOI: 10.1109/TPEL.2012.2227502. 61

[37] H. Krishnaswami, "Photovoltaic microinverter using single-stage isolated high-frequency link series resonant topology," in *Proc. IEEE ECCE*, 2011, vol. 1, no. 1, pp. 495–500. DOI: 10.1109/ECCE.2011.6063810. 61

[38] K.C.A. de Souza, M.R. De Castro, and F. Antunes, "A DC / AC converter for single-phase grid-connected photovoltaic systems," in *Proc. IEEE IECON*, 2002, pp. 3268–3273. DOI: 10.1109/IECON.2002.1182922. 62

[39] R. Teodorescu, F. Blaabjerg, M. Liserre, and P.C. Loh, "Proportional-resonant controllers and filters for grid-connected voltage-source converters," *IEE Proceedings-Electric Power Appl.*, vol. 150, no. 5, pp. 139–145, 2003. 63

[40] Z. Liang, R. Guo, J. Li, and A.Q. Huang, "A High-efficiency PV module-integrated DC/DC converter for PV energy harvest in FREEDM systems," *IEEE Trans. Power Electron.*, vol. 26, no. 3, pp. 897–909, 2011. DOI: 10.1109/TPEL.2011.2107581. 65

[41] S.-J. Jang, Chung-YuenWon, B.-K. Lee, and J. Hur, "Fuel cell generation system with a new active clamping current-fed half-bridge converter," *IEEE Trans. Energy Convers.*, vol. 22, no. 2, pp. 332–340, 2007. DOI: 10.1109/TEC.2006.874208. 66

[42] D.R. Nayanasiri, D.M. Vilathgamuwa, and D.L. Maskell, "Optimized switching Control strategy for current-fed half-bridge converter," in *Proc. IEEE APEC*, 2014, pp. 2023–2028. DOI: 10.1109/APEC.2014.6803584. 66

[43] W. de A. Filho and I. Barbi, "A comparison between two current-fed push-pull DC-DC converters - analysis, design and experimentation," in *Proc. IEEE INTELEC*, 1996, vol. 00, pp. 313–320. DOI: 10.1109/INTLEC.1996.573330. 72

[44] N.A. Ninad and L.A.C. Lopes, "A low power single -phase utility interactive inverter for residential PV generation with small DC link capacitor." 83

[45] S.A. Khajehoddin, M.K. Ghartemani, P.K. Jain, and A. Bakhshai, "DC-bus design and control for a single phase grid-connected renewable converter with small energy storage component," *IEEE Trans. Power Electron.*, vol. 28, no. 7, pp. 3245–3254, 2011. DOI: 10.1109/TPEL.2012.2222449. 83

[46] T. Hirao, T. Shimizu, M. Ishikawa, and K. Yasui, "A modified modulation control of a single-phase inverter with enhanced power decoupling for a photovoltaic AC module," in *European Conference on Power Electronics and Applications*, 2005. DOI: 10.1109/EPE.2005.219387. 85

[47] B.J. Pierquet and D.J. Perreault, "A single-phase photovoltaic inverter topology with a series-connected power buffer," in *Proc. IEEE ECCE*, 2010, pp. 2811–2818. DOI: 10.1109/TPEL.2013.2237790. 86

[48] Q. Li, P. Wolfs, and S. Senini, "A hard switched high frequency link converter with constant power output for photovoltaic applications," in *Proc. ACPE*, 2002, pp. 1–6. 88

[49] R. Kathiresan, P. Das, T. Reindl, and S.K. Panda, "Purely inductive ripple power storage for improved lifetime in solar photovoltaic micro-inverter topology," in *Proc. IEEE PVSC*, 2014, pp. 980–985. DOI: 10.1109/PVSC.2014.6925079. 89

[50] R.-L. Lin and C.-W. Lin, "Design criteria for resonant tank of LLC DC-DC resonant converter," in *Proc. IEEE Industrial Electronics Society Annual Conference IECON 2010*, pp. 427, 432, 7-10 Nov. 2010. 110, 111

Authors' Biographies

MAHINDA VILATHGAMUWA

Mahinda Vilathgamuwa obtained his B.Sc. and Ph.D. degrees from University of Moratuwaand University of Cambridge in 1984 and 1988, respectively. In 1985 he started his academic career as an assistant lecturer at University of Moratuwa. Later, after obtaining a Ph.D. in Electrical Engineering from University of Cambridge, England, he became a Senior Lecturer at the same University. Since 1993 Mahinda served as an academic in the capacities of Lecturer, Assistant Professor, and Associate Professor at Nanyang Technological University in Singapore. In 2014 he joined the Queensland University of Technology in Brisbane Australia where he is currently a Professor of Electrical Engineering and Computer Science. Mahinda is also a Senior Member of Institute of Electrical and Electronics Engineers.

DULIKA NAYANASIRI

Dulika Nayanasiri received his B.Sc. degree in Electronics and Telecommunication Engineering from University of Moratuwa, Moratuwa, Sri Lanka, in 2010 and his Ph.D. degree in Electrical Engineering from Nanyang Technological University, Singapore in 2015. Currently, he is working as a lecturer in the Electronics and Telecommunications Engineering department at University of Moratuwa. His research interests include power electronic converters and their application in renewable energy, especially in grid-connected photovoltaic systems.

SHANTHA GAMINI

Shantha Gamini received his B.Sc. degree in Electronics and Telecommunication Engineering from University of Moratuwa, Sri Lanka, in 2003, and his Ph.D. degree in Electrical Engineering from Nanyang Technological University, Singapore in 2013. From 2011–2015 he worked as an Electrical Systems Engineer at Rolls Royce Advanced Technology Centre in Singapore. Currently, he is a lecturer in maritime electrical engineering at the Australian Maritime College at University of Tasmania, Australia. His research interests include power electronic converters, renewable energy technologies, grid integration of energy systems, shipboard power systems, and electric propulsion. Dr. Shantha has published over 30 scientific papers in international journals and conference proceedings.

Printed in the United States
by Baker & Taylor Publisher Services